GRAPHING CALCULATOR MANUAL

KATHLEEN MCLAUGHLIN
University of Connecticut

DOROTHY WAKEFIELD
University of Connecticut Health Center

ELEMENTARY STATISTICS: PICTURING THE WORLD

FIFTH EDITION

Ron Larson
Pennsylvania State University

Betsy Farber
Bucks County Community College

Prentice Hall
is an imprint of

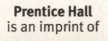

PEARSON

Reproduced by Pearson Prentice Hall from electronic files supplied by the author.

Copyright © 2012, 2009, 2006 Pearson Education, Inc.
Publishing as Prentice Hall, 75 Arlington Street, Boston, MA 02116.

ISBN-13: 978-0-321-69379-2
ISBN-10: 0-321-69379-5

1 2 3 4 5 6 BRR 15 14 13 12 11

Prentice Hall
is an imprint of

www.pearsonhighered.com

▶ Contents

▶ Introduction

The TI-83/TI-84 Graphing Calculator Manual is one of a series of companion technology manuals that provide hands-on technology assistance to users of Larson/Farber *Elementary Statistics: Picturing the World,* 5th Edition.

Detailed instructions for working selected examples, exercises, and technology sections from *Elementary Statistics: Picturing the World,* 5[th]Ed. are provided in this manual. To make the correlation with the text as seamless as possible, the table of contents includes page references for both the Larson/Farber text and this manual.

Getting Started with theTI-83 and TI-84 Graphing Calculators

▶ Overview

This manual is designed to be used with the TI-83 and TI-84 families of Graphing Calculators. These calculators have a variety of useful functions for doing statistical calculations and for creating statistical plots. The commands for using the statistical functions are basically the same for the TI-83s and TI-84s. All TI-84 calculators, the TI-83 Plus Calculator, and the TI-83 Silver Edition can receive a variety of software applications that are available through the TI website (www.ti.com). TI also will provide downloadable updates to the operating systems of these calculators. These features are not available on the TI-83.

Your textbook comes with data files on the CD data disk that can be loaded onto any of the TI-83 or TI-84 calculators. The requirements for the transfer of data differ from one calculator to another. Some of these calculators (TI-84s and TI-83 Silver Edition) are sold with the necessary connection. For the TI-83 or TI-83 Plus, you can purchase a Graph Link manufactured by Texas Instruments which connects the calculator to the computer. (Note: In order to do examples in this manual, you can simply enter the data values for each example directly into your calculator. It is not necessary to use the graph link to download the data into your calculator. The download procedure using the computer link is an optional way of entering data.)

Throughout this manual all instructions and screen shots use the TI-84/Plus. These instructions and screen shots are compatible with all the TI-84 and TI-83 calculators.

Before you begin using the TI-83 or TI-84 calculator, spend a few minutes becoming familiar with its basic operations. First, notice the different colored keys on the calculator. On the TI-84s, the white keys are the number keys; the light gray keys on the right are the basic mathematical functions; the dark gray keys on the left are additional mathematical functions; the remaining dark gray keys are the advanced functions; the light gray keys just below the viewing screen are used to set up and display graphs, and the light gray arrow keys are used for moving the cursor around the viewing screen. On the TI-83, the white keys are the number keys; the blue keys on the right are the basic mathematical functions; the dark gray keys on the left are additional mathematical functions; the remaining dark gray keys are the advanced functions; the blue keys just below the viewing screen are used to set up and display graphs, and the blue arrow keys are used for moving the cursor around the viewing screen.

The primary function of each key is printed in white on the key. For example, when you press **STAT**, the STAT MENU is displayed.

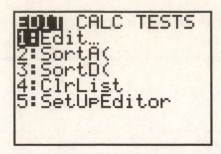

The secondary function of each key is printed in blue on the TI-84 (yellow on the TI-83) above the key. When you press the **2**[nd] key (found in the upper left corner of the keys), the function printed above the key becomes active and the cursor changes from a solid rectangle to an ⬆ (up-arrow). For example, when you press **2**[nd] and the $\boxed{x^2}$ key, the $\sqrt{}$ function is activated.

The notation used in this manual to indicate a secondary function is '**2**[nd]' followed by the name of the secondary function. For example, to use the LIST function, found above the **STAT** key, the notation used in this manual is **2**[nd] **[LIST]**. The LIST MENU will then be activated and displayed on the screen.

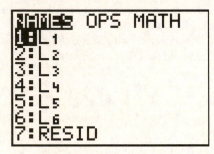

The alpha function of each key is printed in green above the key. When you press the green **ALPHA** key, the function printed in green above the key is activated and the cursor changes from a solid rectangle to **A**.

In this manual you will find detailed explanations of the different statistical functions that are programmed into the TI-83 and TI-84 graphing calculators. These explanations will accompany selected examples from your textbook. This will give you the opportunity to learn the various calculator functions as they apply to the specific statistical material in each chapter.

▶ Getting Started

To operate the calculator, press **ON** in the lower left corner of the calculator. Begin each example with a blank screen, with a rectangular cursor flashing in the upper left corner. If you turn on your calculator and you do not have a blank screen, press the **CLEAR** key. You may have to press **CLEAR** a second time in order to clear the screen. If using the **CLEAR** key does not clear the screen, you can push **2**[nd] **[QUIT]** (Note: **QUIT** is found above the **MODE** key.)

▶ Helpful Hints

To adjust the display contrast, push and release the 2nd key. Then push and hold the up arrow ▲ to darken or the down arrow ▼ to lighten.

The calculator has an automatic turn off that will turn the calculator off if it has been idle for several minutes. To restart, simply press the **ON** key.

There are several different graphing techniques available on the TI-83 and TI-84 calculators. If you inadvertently leave a graph on and attempt to use a different graphing function, your graph display may be cluttered with extraneous graphs, or you may get an ERROR message on the screen.

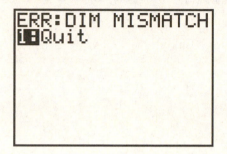

There are several items that you should check before graphing anything. First, press the **Y=** key, found in the upper left corner of the key pad, and clear all the Y-variables. The screen should look like the following display:

If there are any functions stored in the Y-variables, simply move the cursor to the line that contains a function and press **CLEAR** **ENTER**.

Next, press 2nd **[STAT PLOT]** (found on the **Y=** key) and check to make sure that all the STAT PLOTS are turned **OFF**.

If you notice that a Plot is turned **ON**, select the Plot by using the down arrow key to highlight the number to the left of the Plot, press **ENTER** and move the cursor to **OFF** and press **ENTER**. Press **2**[nd] **[QUIT]** to return to the home screen.

Now you are ready to get started with your calculator.   Enjoy!!

Introduction to Statistics

▶ Technology (pg. 34 - 35) Example: Generating A List of Random Numbers

The first step is to initialize your calculator to generate random numbers by setting a unique starting value called a *seed*. You set the *seed* by selecting any 'starting number' and storing this number in **rand**. Suppose, for this example, that we select the number '34' as the starting number. Type **34** into your calculator and press the **STO** key found in the lower left section of the calculator keys. Next press the **MATH** key found in the upper left section of the calculator keys. The Math Menu will appear.

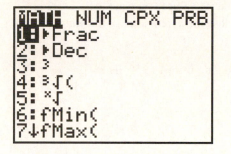

Use the right arrow key ▶ found in the upper right section of the calculator keys, to move the cursor to highlight **PRB**. The Probability Menu will appear.

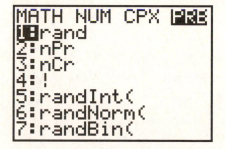

The first selection on the **PRB** menu is **rand,** which stands for 'random number.' Notice that this highlighted. Simply press **ENTER** twice and the starting value of '34' will be stored into **rand** and will be used as the *seed* for generating random numbers. (Note: This example uses '34' as the *seed,* but you should use your own number as a seed for your random number generator. You only need to do this initialization process once. You do not need to do this every time you generate random numbers.)

```
34→rand
              34
■
```

 Now you are ready to generate random numbers.

To generate a random sample of integers, press **MATH**. Use the right arrow key ▶ to move the cursor to highlight **PRB**. The Probability Menu will appear. Select **5:RandInt(** by using the down arrow key ▼ to highlight it and pressing **ENTER** or by pressing the **5** key. **RandInt(** should appear on the screen. This function requires three values: the starting integer, followed by a comma (the comma is found on the black key above the **7** key), the ending integer, followed by a comma and the number of values you want to generate. Close the parentheses and press **ENTER**. (Note: It is optional to close the parentheses at the end of the command.)

For an example, suppose you want to generate 15 values from the integers ranging from 1 to 50. The command is **randInt(1,50,15)**.

```
randInt(1,50,15)
■
```

Press **ENTER** and a partial display of the 15 random integers should appear on your screen. (Note: Your numbers will be different from the ones you see here.)

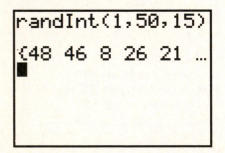

Use the right arrow to scroll through your 15 items. You might find that you have some duplicate values. The TI-83/84 Plus uses a method called "sampling with replacement" to generate random numbers. This means that it is possible to select the same integer twice.

In the example in the text, you are asked to select a random sample of 15 cars from the 167 cars that are assembled at an auto plant. One way to choose the sample is to number the cars from 1 to 167 and randomly select 15 different cars. This sampling process is to be "without replacement." Since the TI-83/84 Plus samples "with replacement" the best way to obtain 15 different cars is to generate more than 15 random integers, and to discard any duplicates. To be safe, you should generate 20 random integers.

```
randInt(1,167,20
)
{146 119 40 81 …
■
```

Use the right arrow to scroll to the right to see the rest of the list and write down the first 15 distinct values.

Exercises

1. You would like to sample 10 distinct brokers. The TI-83/84 Plus samples "with replacement" so you may have duplicates in your sample. To obtain 10 distinct values, try selecting 15 random numbers. Press **MATH**, highlight **PRB** and select **5:RandInt(**. Press **ENTER** and type in **1 , 86 , 15)**. Press **ENTER** and the random integers will appear on the screen. Use the right arrow to scroll through the output and choose the first ten distinct integers. (Note: If you don't have 10 distinct integers, simply press **ENTER** to generate 15 more integers and pick as many new integers as you need from this group to complete your sample of 10).

2. You would like to sample 25 distinct camera phones. To obtain 25 distinct values, try selecting 30 random numbers. Press **MATH**, highlight **PRB** and select **5:RandInt(**. Press **ENTER** and type in **1 , 300 , 30)** and press **ENTER**. The random integers will appear on the screen. Use the right arrow to scroll through the output and choose the first twenty-five distinct integers. (Note: If you don't have 25 distinct integers, simply press **ENTER** to generate 30 more integers and pick as many new integers as you need from this group to complete your sample of 25).

3. This example does not require distinct digits, so duplicates are allowed. You can select three random samples of size n=5 by using **randInt** to create each sample. To generate the first sample, press **MATH**, highlight **PRB** and select **5:RandInt(** by scrolling down through the list and highlighting **5:RandInt(** and pressing **ENTER** or by highlighting **PRB** and pressing **5**. Type in **0 , 9 , 5.** Close the parentheses and press **ENTER**.

```
randInt(0,9,5)
      {7 3 2 9 9}
■
```

Add the digits and divide by 5 to get the average for the sample.

To generate the second and third samples, simply press **ENTER** twice. Calculate the sample average for the 2nd and 3rd sample. To calculate the population average, add the digits 0 through 9 and divide the answer by 10. Compare your three sample averages with your population average.

4.   Use the same procedure as in Exercise 3 with a starting value of 0 and an ending value of 40 and a sample size of 7.

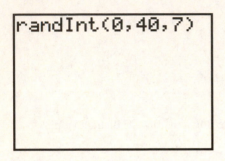

5.   In this example, you need to generate the random data, store the data in a list and then sort the data. To enter the data, press **STAT** and the Statistics Menu will appear. Notice that **EDIT** is highlighted.

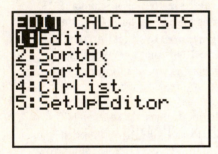

Press **ENTER** and lists **L1**, **L2** and **L3** will appear.

If the lists already contain data, you should clear them before beginning this example. To clear data from a list, move your cursor so that the List name (**L1**, **L2**, or **L3**) for the list that contains data is highlighted.

```
 L1    |L2    |L3       1
 1     |4     |------
 2     |6     |
 3     |8     |
 10    |      |
 ------|------|------

L1 = {1,2,3,10}
```

Press **CLEAR** **ENTER**. Repeat this process to empty data stored in **L2** and **L3**.

Move the cursor to the top of L1 so that **L1**, the listname, is highlighted and press **ENTER**. The cursor will now be flashing at the bottom of the screen after "**L1=**." Press **MATH**, highlight **PRB** and select **5:RandInt(** by scrolling down through the list and highlighting **5:RandInt(** and pressing **ENTER** or by highlighting **PRB** and pressing **5**. Type in **1** , **6** , **60.** Close the parentheses and press **ENTER**.

```
 L1    |L2    |L3       1
 ------|------|------




L1 = ...Int(1,6,60)
```

Next, sort L1 by pressing **2ⁿᵈ** **[LIST]** (**[LIST]** is found above the **STAT** key), highlight **OPS** and select **1:SortA(** and press **2ⁿᵈ** **[L1]** Close the parentheses and press **ENTER**. Press **2ⁿᵈ** **[L1]** **ENTER** and a partial list of the integers in L1 will appear on the screen. Use the right arrow to scroll through the list and count the number of 1's, 2's, 3's, … and 6's. Record the results in a table.

7. Use the same procedure as in Exercise 5 with a starting value of 0, an ending value of 1 and a sample size of 100. After you have generated the values in L1, press **2ⁿᵈ** **[QUIT]** (**[QUIT]** is found above the **MODE** key). Press **2ⁿᵈ** **[L1]** highlight **MATH** and select **5:sum(** and press **2ⁿᵈ** **[L1]**. Close the parentheses and press **ENTER**.

```
sum(L1)
                44
```

The number in your output is the sum of L1. Since L1 consists of 0's and 1's, the sum is actually the number of tails in your random sample. The (number of heads) = 100 - (no. of tails). (Note: Your simulation results will differ from those shown in this example.)

Descriptive Statistics

Section 2.1

▸ Example 7 (pg. 46) Constructing a Histogram for a Frequency Distribution

To create this histogram, you must enter information into List1 (**L1**) and List 2 (**L2**). Refer to the frequency distribution on pg. 41 in your textbook. You will enter the midpoints into **L1** and the frequencies into **L2**. To enter the data, press **STAT** and the Statistics Menu will appear. Notice that **EDIT** is highlighted. The first selection in the **EDIT** menu, **1:Edit**, is also highlighted.

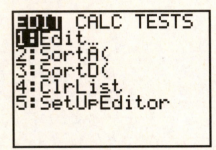

Press **ENTER** and lists **L1**, **L2** and **L3** will appear.

```
L1      L2      L3      2

------  ██████  ------




L2(1)=
```

If the lists already contain data, you should clear them before beginning this example. To clear data from a list, move your cursor so that the List name (**L1**, **L2**, or **L3**) for the list that contains data is highlighted.

```
┌─────────┬─────────┬─────────┬───┐
│ ▌1      │ L2      │ L3      │ 1 │
├─────────┼─────────┼─────────┼───┤
│ 1       │ 4       │ ------  │   │
│ 2       │ 6       │         │   │
│ 3       │ 8       │         │   │
│ 10      │         │         │   │
│ ------  │         │         │   │
├─────────┴─────────┴─────────┴───┤
│ L1 ={1,2,3,10}                  │
└─────────────────────────────────┘
```

Press **CLEAR** **ENTER**. Repeat this process to empty data stored in L2 and L3.

```
┌─────────┬─────────┬─────────┬───┐
│ L1      │ L2      │ L3      │ 1 │
├─────────┼─────────┼─────────┼───┤
│ ▬▬▬▬▬▬  │ 4       │ ------  │   │
│         │ 6       │         │   │
│         │ 8       │         │   │
│         │ ------  │         │   │
├─────────┴─────────┴─────────┴───┤
│ L1(1)=                          │
└─────────────────────────────────┘
```

To enter the midpoints into **L1,** move your cursor so that it is positioned in the 1st position in **L1**. Type in the first midpoint, **86.5,** and press **ENTER** or use the **Down Arrow**. Enter the next midpoint, **142.5.** Continue this process until all 7 midpoints are entered into **L1**. Now use the **Up Arrow** to scroll to the top of **L1**. As you scroll through the data, check it. If a data point is incorrect, simply move the cursor to highlight it and type in the correct value. When you have moved to the 1st value in **L1**, use the right arrow to move to the first position in **L2**. Enter the frequencies into **L2**.

```
┌─────────┬─────────┬─────────┬───┐
│ L1      │ L2      │ L3      │ 2 │
├─────────┼─────────┼─────────┼───┤
│ 86.5    │ 5       │ ------  │   │
│ 142.5   │ 8       │         │   │
│ 198.5   │ 6       │         │   │
│ 254.5   │ 5       │         │   │
│ 310.5   │ 2       │         │   │
│ 366.5   │ 1       │         │   │
│ 422.5   │ ▮3▮     │         │   │
├─────────┴─────────┴─────────┴───┤
│ L2(7) =3                        │
└─────────────────────────────────┘
```

To prepare to construct the histogram, press the **Y=** key and clear all the Y-registers. To graph the histogram, press **2nd** [STAT PLOT] (located above the **Y=** key).

Select **Plot1** by pressing ENTER.

Notice that **Plot1** is highlighted. On the next line, notice that the cursor is flashing on **On** or **Off**. Position the cursor on **On** and press ENTER to select it. The next two lines on the screen show the different types of graphs. Move your cursor to the symbol for histogram (3rd item in the 1st line of **Type**) and press ENTER.

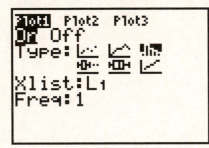

The next line is **Xlist**. Use the **Down Arrow** to move to this line. On this line, you tell the calculator where the data (the midpoints) are stored. In most graphing situations, the data are entered into **L1** so **L1** is the default option. Notice that the cursor is flashing on **L1**. Push ENTER to select **L1**. The last line is the frequency line. On this line, **1** is the default. The cursor should be flashing on **1**. Change **1** to **L2** by pressing 2nd [**L2**].

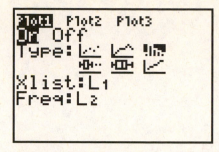

To view a histogram of the data, press **ZOOM**.

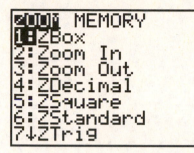

There are several options in the Zoom Menu. Using the **Down Arrow**, scroll down to option **9**, **ZoomStat,** and press **ENTER**. A histogram should appear on the screen.

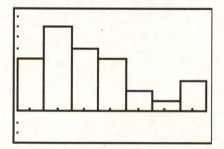

If your histogram does not look exactly like the one above (which is the same as the one on pg. 46 of your textbook), you can make adjustments to your histogram. Press **WINDOW** and set **Xmin** to 86.5, **Xmax** to 534.5 (this one extra midpoint is recommended to make sure that the complete histogram appears on the screen), and **Xscl** to 56, which is the difference between successive midpoints in the frequency distribution. Note: It is not necessary for you to make any changes to **Ymin**, **Ymax** or **Yscl**.

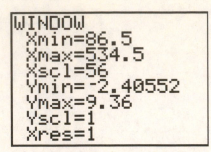

Press **GRAPH**.

Notice the **TRACE** key (located just below the output screen). If you press it, a flashing cursor, *, will appear at the top of the 1st bar of the histogram.

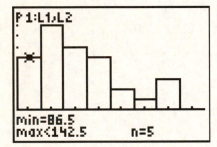

The value of the smallest midpoint is displayed as 86.5 and the number of data points in that bar is displayed as n = 5. Use the **Right Arrow** to move through each of the bars.

Now that you have completed this example, turn Plot1 **Off**. Using **2nd** [STAT PLOT], select **Plot1** by pressing **ENTER** and highlighting **Off**. Press **ENTER** and **2nd** [QUIT]. (Note: Turning Plot1 **Off** is optional. You can leave it **On** but leaving it **On** will affect other graphing operations.)

◀

▶ Exercise 31 (pg. 50) Construct a Frequency Histogram Using 6 Classes

Press **STAT** and **ENTER** to select **1:Edit**. Highlight **L1** at the top of the first list and press **CLEAR** and **ENTER** to clear the data in **L1**. You can also clear **L2** although you will not be using **L2** in this example. Enter the data into **L1**. Scroll through your completed list and verify each entry.

```
L1          L2      L3       1
█████     ------  ------
2468
7119
1876
4105
3183
1932
L1(1)=2114
```

To prepare to construct the histogram, press the **Y=** key and clear all the Y-registers. To set up the histogram, push **2ⁿᵈ [STAT PLOT]** and **ENTER** to select **Plot 1**. Turn **On Plot 1**, set **Type** to **Histogram**, set **Xlist** to **L1**. In this example, you must set **Freq** to **1**. If the frequency is set on **L2** move the cursor so that it is flashing on **L2** and press **CLEAR**. The cursor is now in ALPHA mode (notice that there is an "A" flashing in the cursor). Push the **ALPHA** key and the cursor should return to a solid flashing square. Type in the number **1**.

```
Plot1   Plot2   Plot3
On  Off
Type:  ⌐⌐  ⌐⌐  ▥
       ⊞⋯  ⊞⊞  ⌐⌐
Xlist:L1
Freq:1
```

Press **ZOOM** and scroll down to **9:ZoomStat** and press **ENTER** and a histogram will appear on the screen. Press **WINDOW** to adjust the Graph Window. Set **Xmin** equal to 1000 (the smallest data value) and **Xmax** equal to 7120 (a value, rounded to the nearest ten, that is larger than the largest data point in the dataset). To set the scale so that you will have 6 classes, calculate (**Xmax - Xmin**)/6: (7120 – 1000)/6 = 1020. Use this for **Xscl**. (Note: You do not need to change the values for **Ymin**, **Ymax** or **Yscl**.)

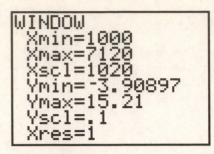

Press **GRAPH** and the histogram should appear.

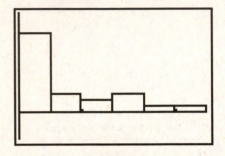

You can press **TRACE** and scroll through the bars of the histogram.
Min and Max values for each bar will appear along with the number of data points in each class.

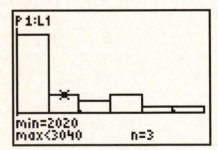

Notice, for example, with the cursor highlighting the second bar of the histogram, you will see that **n=3** appears, indicating that there are 3 data points in the second class which contains values from 2020 to 3039.

Note: After completing a graph, you should turn the graph **Off**. Using 2nd [STAT PLOT], select Plot1 by pressing **ENTER** and highlighting **Off**. Press **ENTER** and 2nd [QUIT]. (Note: Turning Plot1 **Off** is optional. You can leave it **On** but leaving it **On** will affect other graphing operations.)

◀

Section 2.2

▸ Example 5 (pg. 57) Constructing a Pareto Chart

To construct a Pareto chart to represent the causes of inventory shrinkage, you must enter numerical labels for the specific causes into **L1** and the corresponding costs into **L2**. Since the bars are positioned in descending order in a Pareto chart, the labels in **L1** will represent the causes of inventory shrinkage from the most costly to the least costly. Press **STAT** and press **ENTER** to select **1:Edit.** Highlight the name, **L1**, and press **CLEAR** and **ENTER**. Enter the numbers 1, 2, 3 and 4 into **L1**. These numbers represent the four causes of inventory shrinkage: 1 = employee theft, 2 = shoplifting, 3 = administrative error and 4 = vendor fraud. Move your cursor to highlight **L2** and press **CLEAR** and **ENTER**. Enter the corresponding costs: 15.9, 12.7, 5.4 and 1.4.

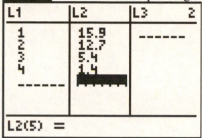

To prepare to construct the Pareto chart, press the **Y=** key and clear all the Y-registers. To draw the Pareto chart, press **2nd** [STAT PLOT]. Press **ENTER** and set up **Plot 1**. Highlight **On** and press **ENTER**. Highlight the histogram icon for **Type** and press **ENTER**. Set **Xlist** to **L1** and **Freq** to **L2**.

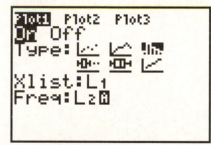

Press **ZOOM** and scroll down to **9:ZoomStat** and press **ENTER** and a histogram will appear on the screen. To create the Pareto chart with unconnected bars, press **WINDOW** and set Xmin = 1 (smallest value in **L1**), Xmax = 5 (one more than the largest value in **L1**) and set Xscl = 0.5. (Note: You do not need to change the values for **Ymin**, **Ymax** or **Yscl**.) Press **GRAPH** to view the Pareto Chart.

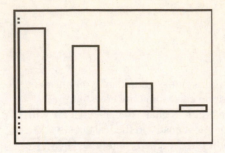

If you press **TRACE** notice, for the first bar, min=1, max <1.5 and n = 15.9. The first bar represents the No. 1 cause of inventory shrinkage. N = 15.9 is the frequency (in millions of dollars) for cause #1. You can use the Right Arrow key to scroll through the remaining bars.

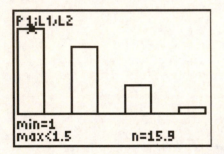

Note: After completing a graph, you should turn the graph **Off**. Using **2ⁿᵈ** [STAT PLOT], select Plot1 by pressing **ENTER** and highlighting **Off**. Press **ENTER** and **2ⁿᵈ** [QUIT]. (Note: Turning Plot1 **Off** is optional. You can leave it **On** but leaving it **On** will affect other graphing operations.)

▶ Try It Yourself 6 (pg. 58)   Constructing a Scatterplot

Press **STAT** and select **1:Edit** from the **Edit menu**. Clear the lists and enter the data points for "Length of employment" into **L1** and the data points for "Salary" into **L2**.

To prepare to construct the scatterplot, press the **Y=** key and clear all the Y-registers. To construct the scatterplot, press **2nd** [STAT PLOT] and select **1:Plot 1** and **ENTER**. Turn On **Plot 1**. Set the **Type** to **Scatterplot** which is the 1st icon in the **Type** choices. For **Xlist** select **L1** and for **Ylist** select **L2**. For **Marks** use the first choice, the small square. Press **ZOOM** and scroll down to **9:ZoomStat** and press **ENTER** or simply press **9** and **ZoomStat** will automatically be selected. The graph should appear on the screen.

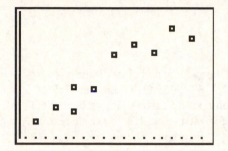

As you can see from the scatterplot, as the length of employment increases, the salary tends to increase.

Note: After completing a graph, you should turn the graph **Off**. Using **2nd** [STAT PLOT], select Plot1 by pressing **ENTER** and highlighting **Off**. Press **ENTER** and **2nd** [QUIT]. (Note: Turning Plot1 **Off** is optional. You can leave it **On** but leaving it **On** will affect other graphing operations.)

◀

▸ Example 7 (pg. 59) Constructing a Time Series Chart

Press **STAT** and select **1:Edit** from the **Edit Menu**. Clear **L1** and **L2**. Enter the "years" into **L1** and the "subscribers" into **L2**.

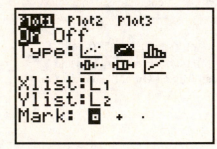

To prepare to construct the time series chart, press the **Y=** key and clear all the Y-registers. To construct the time series chart, press **2ⁿᵈ** [STAT PLOT] and select **1:Plot 1** and **ENTER**. Turn ON **Plot 1**. Set the **Type** to **Connected Scatterplot,** which is the 2ⁿᵈ icon in the **Type** choices. For **Xlist** select **L1** and for **Ylist** select **L2**. Next, there are three different types of **Marks** that you can select for the graph. The first choice, a small square, is the best one to use.

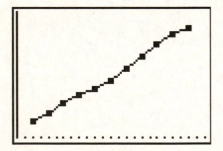

Press **ZOOM** and scroll down to **9:ZoomStat** and press **ENTER** or simply press **9** and **ZoomStat** will automatically be selected. The graph should appear on the screen.

Use **TRACE** and the Right Arrow key to scroll through the data values for each year. Notice for example, the number of subscribers is 140.8 million in 2002.

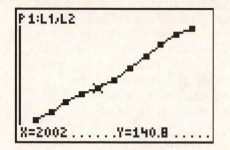

Note: After completing a graph, you should turn the graph **Off**. Using **2ⁿᵈ** [STAT PLOT], select Plot1 by pressing **ENTER** and highlighting **Off**. Press **ENTER** and **2ⁿᵈ** [QUIT]. (Note: Turning Plot1 **Off** is optional. You can leave it **On** but leaving it **On** will affect other graphing operations.)

▸ Exercise 26 (pg. 62)   Pareto Chart

To prepare to construct the Pareto chart, press the **Y=** key and clear all the Y-registers. Press **STAT** and select **1:Edit** from the **Edit Menu**. Clear the lists and enter the numbers 1 through 5 into **L1**. These numbers are labels for the five cities. (Note: 1= Miami, the city with the highest ultraviolet index, 2 = Atlanta, the city with the second highest ultraviolet index, etc.). Enter the ultraviolet indices in descending order into **L2**. Press **2nd** [STAT PLOT] and select **Plot 1** and press **ENTER**. Set the **Type** to **Histogram**. Set **Xlist** to **L1** and Freq to **L2**. Press **ZOOM** and scroll down to **9:ZoomStat** and press **ENTER** and a histogram will appear on the screen.

To create the Pareto chart with unconnected bars, press **WINDOW** and set Xmin = 1 (smallest value in **L1**), Xmax = 6 (1 more than the largest value in **L1**) and set Xscl = 0.5. (Note: You do not need to change the values for **Ymin**, **Ymax** or **Yscl**.) Press **GRAPH** to view the Pareto Chart.

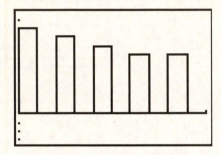

Note: After completing a graph, you should turn the graph **Off**. Using **2nd** [STAT PLOT], select Plot1 by pressing **ENTER** and highlighting **Off**. Press **ENTER** and **2nd** [QUIT]. (Note: Turning Plot1 **Off** is optional. You can leave it **On** but leaving it **On** will affect other graphing operations.)

◂

▶ Exercise 28 (pg. 63) Scatterplot

To prepare to construct the scatterplot, press the **Y=** key and clear all the Y-registers.
Press **STAT** and select **1:Edit** from the **Edit menu**. Clear the lists and enter the data
points for "number of students per teacher" into **L1** and the data points for "average
teacher's salary" into **L2**.

To construct the scatterplot, press **2ⁿᵈ** [STAT PLOT] and select **1:Plot 1** and **ENTER**.
Turn ON **Plot 1**. Set the **Type** to **Scatterplot** which is the 1ˢᵗ icon in the **Type** choices.
For **Xlist** select **L1** and for **Ylist** select **L2**. For **Marks** use the first choice, the small
square. Press **ZOOM** and press **9** for **ZoomStat** and view the graph.

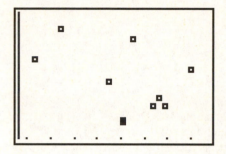

As you can see from the scatterplot, this data set has a large amount of scatter. You
might notice a slight downward trend that suggests that the average teacher's salary
decreases as the number of students per teacher increases.

Note: After completing a graph, you should turn the graph **Off**. Using **2ⁿᵈ** [STAT
PLOT], select Plot1 by pressing **ENTER** and highlighting **Off**. Press **ENTER** and **2ⁿᵈ**
[QUIT]. (Note: Turning Plot1 **Off** is optional. You can leave it **On** but leaving it **On** will
affect other graphing operations.)

◀

▶ Exercise 30 (pg. 63) Time Series Chart

To prepare to construct the Time Series chart, press the **Y=** key and clear all the Y-registers. Press **STAT** and select **1:Edit**. Enter the "years" into **L1** and the "percentages" into **L2**. Press **2nd** [STAT PLOT] and select 1: **Plot 1** and press **ENTER**. Turn **ON Plot 1** and select the **connected scatterplot** (2nd icon) as **Type**. Set **Xlist** to **L1** and **Ylist** to **L2**. For **Marks** use the first choice, the small square. Press **ZOOM** and press **9** for **ZoomStat** and view the graph.

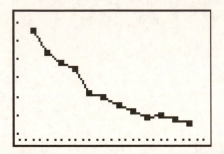

Note: After completing a graph, you should turn the graph **Off**. Using **2nd** [STAT PLOT], select Plot1 by pressing **ENTER** and highlighting **Off**. Press **ENTER** and **2nd** [QUIT]. (Note: Turning Plot1 **Off** is optional. You can leave it **On** but leaving it **On** will affect other graphing operations.)

◀

Section 2.3

▸ Example 6 (pg. 68) Comparing the Mean, Median, and Mode

Press **STAT** and select **1:Edit**. Clear **L1** and enter the data into **L1**. Press **STAT** again
and highlight **CALC** to view the Calc Menu.

Select **1:1-Var Stats** and press **ENTER**. On this line, enter the name of the column that
contains the data. Since you have stored the data in **L1,** simply enter **2ⁿᵈ** **[L1]** **ENTER**
and the first page of the one variable statistics will appear. (Note: If you did not enter a
column name, the default column, which is **L1,** would be automatically selected.)

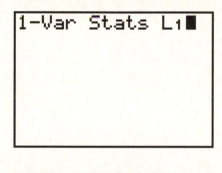

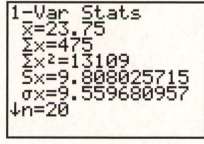

The first item is the mean, $\bar{x} = 23.75$. Notice the down arrow in the bottom left corner of
the screen. This indicates that more information follows this first page. Use the down
arrow key to scroll through this information. The third item you see on the second page
is the median, Med = 21.5.

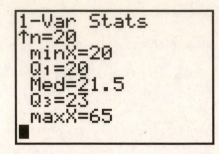

The TI-83/84 Plus does not calculate the mode but, since this data set is sorted, it is easy to see from the list of data that the mode is 20.

▶ Example 7 (pg. 69) Finding a Weighted Mean

Press **STAT** and select **1:Edit**. Clear **L1** and **L2**. Enter the scores into **L1** and the weights into **L2**.

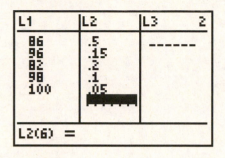

Press **STAT** and highlight **CALC** to view the Calc Menu. Select **1:1-Var Stats**, press **ENTER** and press **2**nd **[L1]** , **2**nd **[L2]**. Press **ENTER**. (Note: You must place the comma between **L1** and **L2**).

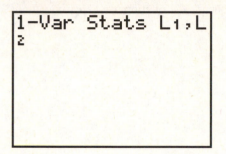

Using **L1** and **L2** in the **1:1-Var Stats** calculation is necessary when calculating a weighted mean. The calculator uses the data in **L1** and the associated weights in **L2** to calculate the average. In this example, the weighted mean is 88.6.

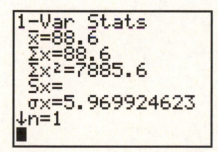

▶ Example 8 (pg. 70) Finding the Mean of a Frequency Distribution

Press **STAT** and select **1:Edit**. Clear **L1** and **L2**. Enter the x-values into **L1** and the frequencies into **L2**. Press **STAT**, highlight **CALC**, select **1:1-Var Stats**, press **ENTER**. Next, press **2ⁿᵈ** [L1] , **2ⁿᵈ** [L2] **ENTER**.

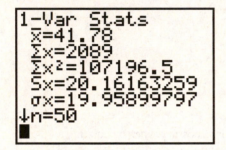

After you press **ENTER**, the sample statistics will appear on the screen. The mean of the frequency distribution described in columns **L1** and **L2** is 41.78. In this example you do not have the actual data. What you have is the frequency distribution of the data summarized into categories. The mean of this frequency distribution is an approximation of the mean of the actual data.

```
1-Var Stats
 x̄=41.78
 Σx=2089
 Σx²=107196.5
 Sx=20.16163259
 σx=19.95899797
↓n=50
■
```

▶ Exercise 21 (pg. 73) Finding the Mean, Median, and Mode

Enter the data into **L1**. Press **STAT** and select **1:1-Var Stats** from the Calc Menu. Press **2ⁿᵈ L1 ENTER**.

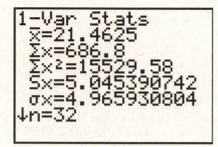

The first item in the output screen is the mean, 21.4625. Scroll down through the output and find the median, Med = 21.95.

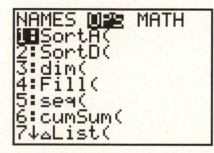

Although the TI-83/84 Plus does not calculate the Mode, you can use the SORT feature to order the data. You can then scroll through the data to see if the data set contains a mode. To sort the data, press **2ⁿᵈ [LIST]**. (Note: **List** is found above the **STAT** key). Move the cursor to highlight **OPS** and select 1:**SortA(** and **ENTER**.

To sort **L1** in ascending order, press **2nd** **[L1]**, close the parentheses and press **ENTER**.

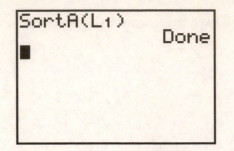

To view the data, press **STAT** and select **1:Edit** and press **ENTER**. Use the down arrow to scroll through **L1** to see if the data has a mode. Look for data values that occur more than once. Record their values and their frequencies. Notice that the value 20.4 occurs twice. All the other values occur only once. The mode is 20.4 for this dataset.

◀

▸ Exercise 51 (pg. 76) Finding the Mean of Grouped Data

Press **STAT** and select **1:Edit**. Clear **L1** and **L2**. Enter the midpoints of each Age group into **L1** and the frequencies into **L2**.

L1	L2	L3 3
4.5	55	
14.5	70	
24.5	35	
34.5	56	
44.5	74	
54.5	42	
64.5	38	
L3(1)=		

Press **STAT** and highlight **CALC** to view the Calc Menu. Select **1:1-Var Stats**, press **ENTER** and type in **2ⁿᵈ** **[L1]** **,** **2ⁿᵈ** **[L2]**. The mean of this grouped data is 35.76.

```
1-Var Stats
 x̄=35.75944584
 Σx=14196.5
 Σx²=698629.25
 Sx=21.96014969
 σx=21.93247463
↓n=397
```

▸ Exercise 55 (pg. 77) Construct a Frequency Histogram

To prepare to construct the histogram, press the **Y=** key and clear all the Y-registers.
Press **STAT** and select **1:Edit**. Clear **L1** and enter the data into **L1**. Press **2**nd [STAT
PLOT] and select **1:Plot 1** and press **ENTER**. Turn **ON** Plot 1 and set the **Type** to
Histogram and press **ENTER**. Move the cursor to **Xlist** and set this to **L1**. Move the
cursor to **Freq** and set it equal to **1**. (Note: The cursor may be in **ALPHA** mode with a
flashing **A**. Press **ALPHA** to return to the solid rectangular cursor and type in **1**.)

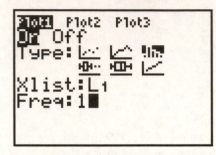

Press **ZOOM** and **9** to select **ZoomStat.** The histogram that is displayed has 7 classes.
To change to 5 classes, press **Window**. Notice that **Xmin** = 62 and **Xmax** = 78.33.
Change **Xmax** to 79, which is the next whole number greater than 78.33. To determine a
value for **Xscl**, you must calculate (**Xmax - Xmin**)/5. Press **2**nd [QUIT] to close out the
Window Menu. (Note: QUIT is found above the **MODE** key). Calculate (79 - 62)/5 and
round the resulting value, 3.4, to 3. Press **Window** and set **Xscl** = 3. Press **GRAPH** and
view the histogram with 5 classes. Notice that one of the bars is too tall to fit completely
on the screen. Press **Window** and set **Ymax** = 9, which is the next whole number greater
than 8.19. Press **GRAPH** again. As you can see from the graph, the histogram appears
to be symmetric.

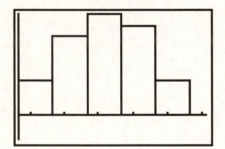

▸ Exercise 64 (pg. 78) Data Analysis

Press **STAT** and select **1:Edit**. Clear **L1** and then enter the data. Press **STAT** and highlight **CALC**. Select **1:1-Var Stats** and press 2nd **L1** **ENTER** to obtain values for the mean and median.

Since the TI-83/84 Plus does not do a stem-and-leaf plot, you can use a histogram to get a picture of the data. To prepare to construct the histogram, press the **Y=** key and clear all the Y-registers. Press 2nd [STAT PLOT], select **1:Plot 1** and press **ENTER**. Turn **On Plot 1**. Set **Type** to **Histogram**. The **Xlist** is **L1** and the **Freq** is **1**. Press **ZOOM** and scroll down to **9:ZoomStat** and press **ENTER** and a histogram will appear on the screen.

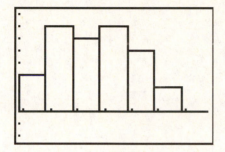

To construct a histogram that would be similar to a stem-and-leaf plot that uses one row per stem, press **WINDOW** and set **Xmin** = 10, **Xmax** = 100 and **Xscl** = 10. Press **GRAPH** to display the new histogram.

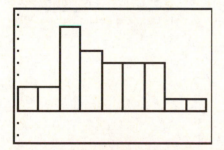

You can press **TRACE** and scroll through the bars of the histogram. Min and Max values for each bar will appear along with the number of data points in each class. You can use this information to determine the location of the mean and the median.

There appears to be some positive skewness in the data. (Note: This is more easily seen in the second of the two histograms.)

◂

Section 2.4

▶ Example 5 (pg. 84) Finding the Standard Deviation

Press **STAT** and select **1:Edit**. Clear **L1** and enter the data into **L1**. Press **STAT** and highlight **CALC** to display the Calc Menu. Select **1: 1-Var Stats** and press **2ⁿᵈ L1** **ENTER**.   The sample standard deviation is Sx, 5.089373342

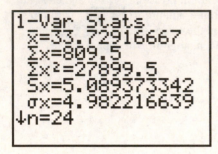

▶ Example 9 (pg. 88) Standard Deviation of Grouped Data

Press **STAT** and select **1:Edit**. Clear **L1** and **L2**. Enter the x-values into **L1** and the frequencies into **L2**. Press **STAT** and highlight **CALC** to select the Calc Menu. Select **1:1-Var Stats** and press **ENTER**. Type in **2ⁿᵈ [L1]** **,** **2ⁿᵈ [L2]** and press **ENTER**. From the statistics displayed, the mean is 1.82 and the sample standard deviation is 1.722.

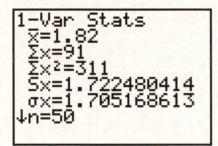

▶ Exercise 25 (pg. 92) Comparing Two Datasets

Press **STAT** and select **1:Edit**. Clear **L1** and **L2**. Enter the Dallas Data into **L1** and the New York City Data into **L2**. Press **STAT** and highlight **CALC**. Select **1:1-Var Stats** and press **2nd** **L1** **ENTER**. For the Dallas data, the sample mean is 44.28, the sample standard deviation is Sx = 4.28 and the sample median (found by scrolling to the 2nd page of output) is 44.7.

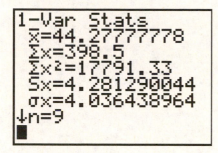

To find the range, scroll down through the display and find **minX.** Continue scrolling through the display and find **maxX.** To calculate the range, simply type in **50 - 38.7** and press **ENTER**. The range = 11.3.

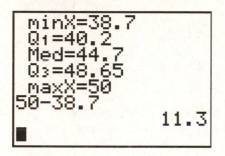

To find the variance, you must square the standard deviation. Type in **4.28** and press the x^2 key. The variance is 18.32.

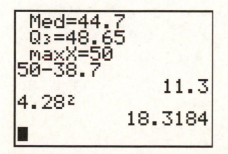

For the New York City Data, press **STAT**, highlight **CALC**, select **1:1-Var Stats**, press **ENTER** and press **2nd** **L2.**

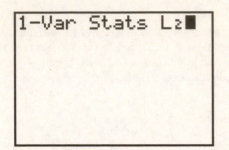

Press **ENTER**. For the New York City data, the sample mean is 50.91, the sample standard deviation is Sx = 7.096 and the sample median (found by scrolling to the 2nd page of output) is 50.6.

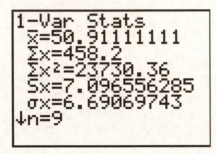

Use the same procedure as you used for the Dallas data, to find the range (17.8) and the variance (50.35) for the New York City data.

The annual salaries in New York City are more variable than the annual salaries in Dallas. The annual salaries in Dallas have a lower mean and a lower median than the annual salaries New York City.

Section 2.5

▶ Example 2 (pg. 101) Finding Quartiles

Press **STAT** and select **1:Edit**. Clear **L1** and enter the data into **L1**. Press **STAT** and highlight **CALC**. Select **1:1-Var Stats** and press **2ⁿᵈ L1 ENTER**. Scroll down through the descriptive statistics. You will see the first quartile: **Q1 = 21.5,** the second quartile (the median): **Med = 23** and the third quartile: **Q3 = 28.**

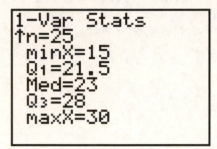

▶ Example 4 (pg. 103) Drawing a Box-and-Whisker-Plot

To prepare to construct the Box-and-Whisker plot, press the **Y=** key and clear all the Y-registers. Press **STAT** and select **1:Edit**. Clear **L1** and enter the data from Example 1 on pg. 100 in your textbook. Press **2ⁿᵈ** [STAT PLOT]. Select **1:Plot 1** and press **ENTER**. Turn On **Plot 1**. Using the right arrow (you cannot use the down arrow to drop to the second line), scroll through the **Type** options and choose the second boxplot which is the middle entry in row 2 of the **TYPE** options. Press **ENTER**. Move to **Xlist** and type in **2ⁿᵈ L1**. Press **ENTER** and move to **Freq**. Set **Freq** to **1.** If **Freq** is set on **L2**, press **ALPHA** to return the cursor to a flashing solid rectangle and type in **1**. Press **ZOOM** and **9** to select **ZoomStat**. The Boxplot will appear on your screen.

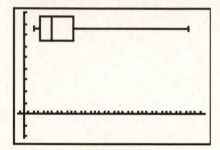

If you press **TRACE** and use the left and right arrow keys, you can display the five values that represent the five-number summary of the data.

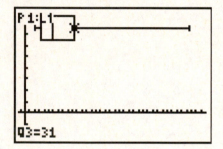

Notice in the above screen display that the trace cursor is on the right side of the box, which represents the third quartile. Also, at the bottom of the screen it is noted that **Q3=31**.

◀

▸ Exercise 23 (pg. 108) Quartiles and a Box-and-Whisker-Plot

To prepare to construct the Box-and-Whisker plot, press the **Y=** key and clear all the Y-registers. Press **STAT** and select **1:Edit**. Clear **L1** and enter the data. Press **2ⁿᵈ** [STAT PLOT]. Select **1:Plot 1** and press **ENTER**. Turn On **Plot 1**. Using the right arrow, scroll through the **Type** options and choose the second boxplot which is the second entry in row 2 of the **TYPE** options. Press **ENTER**. Move to **Xlist** and type in **2ⁿᵈ L1**. Press **ENTER** and move to **Freq**. Set **Freq** to **1.** If **Freq** is set on **L2**, press **ALPHA** to return the cursor to a flashing solid rectangle and type in **1**. Press **ZOOM** and **9** to select **ZoomStat.** The Boxplot will appear on your screen. If you press **TRACE** and use the right and left arrows, you can display **Min, Q1, Med, Q3** and **Max** . Notice **Med**=6 for this example.

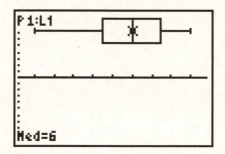

▸ Exercise 31 (pg. 108) Quartiles and a Box-and-Whisker-Plot

To prepare to construct the Box-and-Whisker plot, press the **Y=** key and clear all the Y-registers. Press **STAT** and select **1:Edit**. Clear **L1** and enter the data. Press **2nd [STAT PLOT]**. Select **1:Plot 1** and press **ENTER**. Turn On **Plot 1**. Using the right arrow, scroll through the **Type** options and choose the second boxplot which is the second entry in row 2 of the **TYPE** options. Press **ENTER**. Move to **Xlist** and type in **2nd L1**. Press **ENTER** and move to **Freq**. Set **Freq** to **1**. If **Freq** is set on **L2**, press **ALPHA** to return the cursor to a flashing solid rectangle and type in **1**. Press **ZOOM** and **9** to select **ZoomStat.** The Boxplot will appear on your screen. If you press **TRACE** and use the right and left arrows, you can display **Min, Q1, Med, Q3** and **Max**.

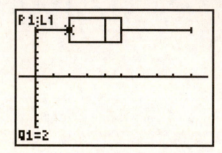

▶ Technology (pg. 121) Monthly Milk Production

Exercises 1-2: Press **STAT** and select **1:Edit**. Clear **L1** and enter the data into **L1**. Press **STAT** and highlight **CALC**. Select **1:1-Var Stats** and press **2ⁿᵈ L1 ENTER**. The sample mean and sample standard deviation can be found in the output display.

Exercises 3-4. You will use the histogram of the data to construct the frequency distribution. To prepare to construct the histogram, press the **Y=** key and clear all the Y-registers. To construct a histogram, press **2ⁿᵈ [STAT PLOT]**. Select **1: Plot 1** and **ENTER**. Turn ON **Plot 1** by highlighting **On** and pressing **ENTER**. Scroll through the **Type** icons, highlight the **Histogram** and press **ENTER**. Move to **Xlist** and type in **2ⁿᵈ L1**. Move to the **Freq** entry and type in **1.** (Note: If the Freq entry is **L2**, Press **ALPHA** and enter **1).** To view a histogram of the data, press **ZOOM** and **9**. To adjust the histogram, press **Window**. Set **Xmin = 1000** and set **Xscl = 500**. (All other entries in the Window Menu can remain unchanged). Press **GRAPH** and then press **TRACE**.

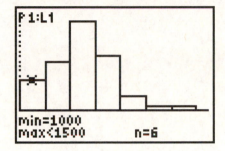

The minimum and maximum values of the first category are displayed. (**1000 and <** **1500**). The midpoint of the first category is (1000+1499)/2, which is **1249.5.** The frequency for the first category is **n = 6.** Trace through the histogram and set up a frequency distribution for the data:

Category	Midpoint	Frequency
1000 – 1499	124 9.5	6

Exercise 5. Using the values that you found in Exercises 1 and 2, calculate the lower and upper endpoints of the one and two standard deviation intervals: $(\bar{x} \pm 1s)$ and $(\bar{x} \pm 2s)$. Next, press **2ⁿᵈ [LIST]**. Highlight **OPS** and select **1:SortA(** , press **ENTER**. Press **2ⁿᵈ [L1]** and close the parentheses. Press **ENTER**. Press **STAT**, select **1:Edit** and press **ENTER**. Scroll through the data in **L1** and count the number of data

points that fall within the one and two standard deviation intervals: $(\overline{x} \pm 1s)$ and $(\overline{x} \pm 2s)$. To find the percentage of data points that lie in each of these intervals, calculate: (number of data points in each interval/ 50) * 100. Compare these percentages with the percentages stated in the Empirical rule.

Exercises 6 - 7. Using the frequency distribution you created, enter the midpoints into **L2** and the frequencies into **L3**. Press **STAT** , highlight **CALC**. Select **1:1-Var Stats** and press **ENTER**. Type in **2nd L2** **,** **2nd L3** and press **ENTER**. The mean and standard deviation for the frequency distribution will be displayed on the screen.

Exercise 8: Compare the actual values for the mean and standard deviation that you found in Exercises 1 and 2 to the estimates for the mean and standard deviation that you found in Exercises 6 and 7.

Probability

Section 3.1

▶ Law of Large Numbers (pg. 134)

You can use the TI83/84 Plus to simulate tossing a coin 150 times. (Note: The scatterplot on pg.134 displays the results of simulating a coin toss 150 times. Your simulation results will differ from the one displayed in this textbook example.)

For this example, you will toss the coin 150 times, store the results in L1 and then sort the results. We will designate "0" as Heads and "1" as Tails.

Press **STAT** and **ENTER** to select **1:Edit**. Highlight **L1** at the top of the first list and press **CLEAR** and **ENTER** to clear the data in **L1**. Move the cursor back to the top of L1, highlight **L1** and press **ENTER**. The cursor will now be flashing next to "L1=" at the bottom of the screen. Press **MATH**, highlight **PRB**, and select **5:randInt(.** The **randInt(** command requires a minimum value, (which is 0 for this simulation), a maximum value (which is 1), and the number of trials (150). In the **randInt(** command type in **0** , **1** , **150**) .

```
┌──────┬──────┬──────┬───┐
│ █1   │ L2   │ L3   │ 1 │
├──────┼──────┼──────┼───┤
│──────│──────│──────│   │
│      │      │      │   │
│      │      │      │   │
│      │      │      │   │
│      │      │      │   │
│      │      │      │   │
├──────┴──────┴──────┴───┤
│ L1 =randInt(0,1,…      │
└────────────────────────┘
```

Press **ENTER**. It will take a few seconds for the calculator to generate 150 tosses. Notice, in the upper right hand corner a flashing ▯ , indicating that the calculator is working. When the simulation has been completed, a string of **0's** and **1's** will appear in L1.

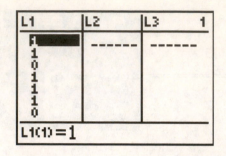

After you have generated the values in L1, press 2nd [QUIT] ([QUIT] is found above the MODE key). Press 2nd [L1], highlight MATH and select 5:sum(and press 2nd [L1]. Close the parentheses and press ENTER.

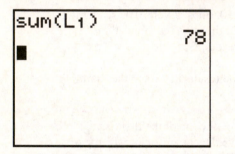

The sum of L1 equals the number of Tails in the list. For this particular simulation, the sum is 78. The number of Heads is equal to (150 - no. of Tails). The proportion of Heads is: (no. of Heads / 150). How close is this proportion to 50%? (Note: Your simulation results will differ from those in this example.)

◄

Section 3.2

▶ Exercise 40 (pg.155) Birthday Problem

To simulate this problem, press **STAT** and **ENTER** to select **1:Edit**. Highlight **L1** at the
top of the first list and press **CLEAR** and **ENTER** to clear the data in **L1**. Move the
cursor back to the top of L1, highlight **L1** and press **ENTER**. The cursor will now be
flashing next to "L1=" at the bottom of the screen. Press **MATH**, highlight **PRB**, and
select **5:randInt(** and enter **1** **,** **365** **,** **24** **)**.

```
L1       L2       L3      1
219      ------   ------
30
92
311
76
26
256
L1(1)=219
```

To see if there are at least two people with the same birthday, you must look for a
matching pair of numbers in **L1**. To do this, press **2ⁿᵈ [LIST]**, highlight **OPS** and select
1:SortA(. The column you want to sort in ascending order is **L1**, so type **2ⁿᵈ [L1])**
into the sort command and press **ENTER**.

Press **STAT** and select **1:EDIT**, press **ENTER** and scroll down through **L1** and check for
matching numbers. If you find any matching numbers that means that at least 2 people in
your simulation have the same birthday.

```
L1       L2       L3      1
15       ------   ------
15
23
44
60
64
74
L1(1)=15
```

Notice in this simulation, 15 is listed twice. This represents two people with the same
birthday, January 15 (the 15ᵗʰ day of a year). That means that for your first simulation

 you have found at least two people with the same birthday. Since you have found this matching pair right at the beginning of **L1**, you do not need to look further into the data in **L1**. Repeat this simulation process nine more times, each time checking to see if you have a matching pair.

How many of your simulations resulted in at least one matching pair? Suppose you found a matching pair in 5 out of your 10 simulations. That means that your estimate of the probability of finding at least 2 people with the same birthday in a room of 24 people is "5 out of 10" or 50%.

◀

Section 3.4

▸ Example 1 (pg. 168) Finding the Number of Permutations

How many different ways can the first row of the Sudoku grid be filled? To find the
number of different arrangements of the numbers 1 through 9, you must calculate the
number of permutations of **9** digits taken **9** at a time. The formula **nPr** is used with **n = 9**
(the number of digits) and **r = 9** (the number of digits you will be selecting). So, the
formula is **9P9**.

Press 9, **MATH**, highlight **PRB** and select **2:nPr** and **ENTER**.

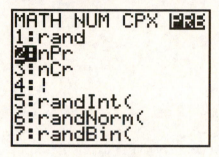

Now press 9 and **ENTER**. The answer, 362880, appears on the screen.

▶ Example 3 (pg. 169)   More Permutations

How many ways can forty-three race cars finish first, second and third in a race? Use the permutation formula with **n = 43** and **r = 3.** Enter **43,** press **MATH**, highlight **PRB**, select **2:nPr** and press **ENTER**. Now enter **3** and press **ENTER**.

```
43 nPr 3
                74046
```

▶ Example 4 (pg. 170)   Distinguishable Permutations

How many distinguishable ways can the 6 one-story houses, 4 two-story houses and 2 split-level houses be arranged? To calculate $\dfrac{12!}{6!4!2!}$ you will use the factorial function

(!). Enter the first value, **12**, press **MATH**, highlight **PRB** and select **4:!.** Then press **÷** . Open the parentheses by pressing **(** . Enter the next value, **6**, press **MATH**, **PRB**, and select **4:!** . To multiply by 4!, press **x** and enter the next value, **4**. Press **MATH**, **PRB**, and select **4:!.** To multiply by 2!, press **x** and enter the next value, **2**. Press **MATH**, **PRB**, and select **4:!.** Close the parentheses **)** and press **ENTER**.

```
12!/(6!*4!*2!)
                13860
```

▶ Example 5 (pg. 171) Finding the Number of Combinations

To calculate the number of different combinations of four companies that can be selected from 16 bidding companies you will use the combination formula: **nCr,** where **n** = the total number of items in the group and **r** = the number of items being selected. In this example, the formula is **16C4**. Enter the first value, **16,** press **MATH**, highlight **PRB** and select **3: nCr,** and enter the second value, **4.** Press **ENTER** and the answer will be displayed on the screen..

```
16 nCr 4
                 1820
```

◀

▶ Example 8 (pg. 173) Finding Probabilities

To calculate the probability of being dealt five diamonds from a standard deck of 52 playing cards, you must calculate: $\dfrac{_{13}C_5}{_{52}C_5}$

To calculate the numerator, enter the first value, **13,** press **MATH**, highlight **PRB** and select **3:nCr** and enter the next value, **5.** Next press **÷** and enter the first value in the denominator, **52,** press **MATH**, highlight **PRB** and select **3:nCr,** enter the next value, **5.** Press **ENTER** and the answer will be displayed on your screen.

```
13 nCr 5/52 nCr
5
    4.951980792E-4
```

Notice that the answer is written in scientific notation. To convert to standard notation, move the decimal point 4 places to the left: 0.000495.

◀

▸ Exercise 54 (pg. 177) Probability

a. To calculate $_{200}C_{15}$, enter the first value, **200**, press **MATH**, highlight **PRB** and select **3:nCr**, enter the next value, **15** and press **ENTER**.

b. Calculate $_{144}C_{15}$ (follow the steps for part a.).

c. The probability that *no* minorities are selected is equal to the probability that the committee is composed completely of non-minorities. This probability can be calculated with the following formula: $\dfrac{_{144}C_{15}}{_{200}C_{15}}$

◂

Technology (pg. 187) Composing Mozart Variations

Exercise 1: The player has 11 phrases to choose from for the 14 bars and 2 phrases for the remaining 2 bars. How many phrases did Mozart write?

Exercise 2: Use the Fundamental Principle of Counting to calculate the number of possible variations. For 14 of the bars there are 11 choices resulting in 11^{14} choices. For the remaining 2 bars, there are 2 choices resulting in 2^2 choices. For the total number of choices, multiply 11^{14} by 2^2.

Exercise 3: To select one number from 1 to 11, press **MATH**, highlight **PRB** and select **5:randInt(** and enter **1** , **11**) and press **ENTER**. One random number between 1 and 11 will be displayed on the screen.

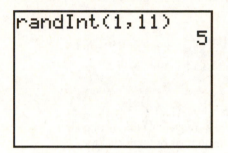

Exercise 3a: Each number from 1 to 11 has an equal chance of being selected. So, each of the 11 possibilities has equal probability. What is the theoretical probability for each number from 1 to 11?

Exercise 3b: To select 100 integers between 1 and 11, press **MATH**, highlight **PRB** and select **5:randInt(** and enter **1** , **11** , **100**) and press **ENTER**. Store the results in **L1** by pressing **STO**, 2nd **[L1]** **ENTER**. Next you can create a histogram and use it to tally your results. To create the histogram, press **2nd** **[STAT PLOT]**, select **1:Plot 1** and press **ENTER**. Turn **ON** Plot 1, set **Type** to **Histogram**. Set **Xlist** to **L1** and **Freq** to **1**. Press **ZOOM** and select **9** for **ZoomStat**. Press **WINDOW** and set **Xscl = 1,** then press **GRAPH**. Use **TRACE** to scroll through the bars of the histogram.

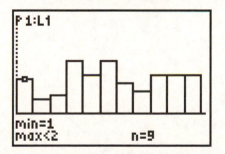

The first bar represents the number of times that a "**1**" occurred in this simulation of the 100 tosses. (Notice on the screen: min = 1 and n = 9). Use the right arrow to scroll through the bars in the histogram. As the cursor moves from one bar in the histogram to the next bar, the "min =" will indicate the current x-value and the "n=" will indicate the corresponding frequency. Record the x-values and the corresponding frequencies in a table. For example, using the histogram in the above diagram, the frequency table would begin with X = 1, n = 9. Convert each frequency into a relative frequency by dividing each frequency by 100. Compare the relative frequencies for each number from 1 to 11 with the theoretical probabilities you obtained in part 3a.

Exercise 4a: The theoretical probability of selecting phrase 6, 7 or 8 for the 1st bar is the sum of the theoretical probabilities you found in part 3a. Raise this answer to the 14th power to find the theoretical probability of selecting phrase 6, 7 or 8 for all 14 bars.

Exercise 4b: Use the relative frequencies found in part 3b. for phrases 6, 7 or 8 to estimate the probability of selecting phrases 6, 7 or 8 for the 1st bar. Raise this answer to the 14th power to estimate the probability of selecting phrase 6, 7 or 8 for all 14 bars.

Exercise 5: To select 2 random numbers from 1 to 6, press **MATH**, highlight **PRB** and select **5:randInt(** and enter **1** **,** **6** **,** **2** **)** and press **ENTER**. Add the two numbers in the display and subtract **1** from the total.

Exercise 5a: First list all possible pairs. (There are 36 pairs: (1,1), (1,2), (1,3)….). For each pair, find the sum and subtract "1" to find a total. Make a frequency table listing all the possible totals from 1 to 11. Next to each total record the number of times that total occurred. Lastly, convert each of these frequencies to relative frequencies by dividing by 36. These relative frequencies are the theoretical probabilities. Use them to answer the question in Exercise 6a.

Exercise 5b: To select 100 totals, create 2 columns of 100 random numbers. Press **MATH**, highlight **PRB** and select **5:randInt(** and enter **1** **,** **6** **,** **100** **)** and press **ENTER**. Press **STO** 2nd **[L1]** and **ENTER** to store this first string of numbers in **L1**. Next create a second set of 100 numbers using the same procedure. Press **STO** 2nd **[L2]** and **ENTER** to store this second string of numbers in **L2**. Now you can calculate the sum of each row and subtract **1**. Press **STAT** and select **1:EDIT**. Move the cursor to highlight **L3**.

L1	L2	L3 3
1	4	------
3	2	
6	4	
2	3	
1	1	
1	2	
3	6	
L3 =		

Press **ENTER** and the cursor will move to the bottom of the screen. On the bottom line, next to **L3 =** type in 2nd **[L1]** + 2nd **[L2]** - 1 and press **ENTER**. The results should appear in **L3**.

L1	L2	L3	3
1	4	4	
3	2	4	
6	4	9	
2	3	4	
1	1	1	
1	2	2	
3	6	8	
L3(1)=4			

Create a histogram of **L3** and use it to tally your results (follow the steps in Exercise 3 to create the histogram). Use **TRACE** to find the frequencies for each X-value and record your results in a frequency table. Convert each frequency to a relative frequency and use the relative frequencies to answer the question in Exercise 6b.

Discrete Probability Distributions

CHAPTER 4

Section 4.1

▶ Example 5 (pg. 194) Mean of a Probability Distribution

Press **STAT** and select **1:EDIT**. Clear **L1** and **L2**. Enter the x-values from the table into **L1** and the P(x) values into **L2**. Press **STAT** and highlight **CALC**. Select **1:1-Var Stats**, press **ENTER** and press 2nd **[L1]** **,** 2nd **[L2]** **ENTER** to see the descriptive statistics. The mean score is 2.94.

```
1-Var Stats
 x̄=2.94
 Σx=2.94
 Σx²=10.26
 Sx=
 σx=1.271377206
↓n=1
```

▶ Example 6 (pg. 195) The Variance and Standard Deviation

This example is a continuation of Example 5. Enter the x-values from the table into **L1** and the P(x) values into **L2**. Press **STAT** and highlight **CALC**. Select **1:1-Var Stats,** press **ENTER** and press 2nd **[L1]** , 2nd **[L2]** **ENTER**. The population standard deviation, σx, is 1.27.

```
1-Var Stats
 x̄=2.94
 Σx=2.94
 Σx²=10.26
 Sx=
 σx=1.271377206
↓n=1
█
```

To find the variance, you must use the value of the standard deviation. Since the variance is equal to the standard deviation squared, type in **1.2714** and press the x^2 key. The population variance is 1.616.

```
Σx=2.94
Σx²=10.26
Sx=
σx=1.271377206
↓n=1
1.2714²
        1.61645796
```

◀

Section 4.2

▶ Example 4 (pg. 206)   Binomial Probabilities

To find a binomial probability you will use the binomial probability density function,
binompdf(n,p,x). For this example, n = 100, p = .67 and x = 75. Press **2ⁿᵈ** **[DISTR]**.
Scroll down through the menu to select **binompdf(** and press **ENTER**. Type in **100** **,**
.67 **,** **75** **)** and press **ENTER**. The answer, 0.0201, will appear on the screen.

```
binompdf(100,.67
,75)
        .0201004116
```

◀

▶ Example 7 (pg. 209) Graphing Binomial Distributions

Construct a probability distribution for a binomial probability model with n = 6 and p = 0.60. Press **STAT**, select **1:EDIT** and clear **L1** and **L2**. Enter the values 0 through 6 into **L1**. Press **2ⁿᵈ** **[QUIT]**. To calculate the probabilities for each X-value in **L1**, first change the display mode so that the probabilities displayed will be rounded to 3 decimal places. Press **MODE** and change from **FLOAT** to **3** press **ENTER** and press **2ⁿᵈ** **[QUIT]**. Next press **2ⁿᵈ** **[DISTR]** and select **binompdf(** and type in **6** **,** **.60** **)** and press **ENTER**. Store these probabilities in **L2** by pressing **STO** **2ⁿᵈ** **[L2]** **ENTER**.

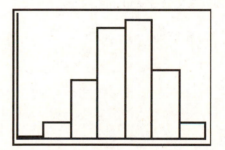

To prepare to construct the probability histogram, press the **Y=** key and clear all the Y-registers. To graph the binomial distribution, press **2ⁿᵈ** **[STAT PLOT]** and press **ENTER**. Turn **ON** Plot 1, select **Histogram** for **Type**, set **Xlist** to **L1** and set **Freq** to **L2**. Press **ZOOM** and select **9** for **ZoomStat.** Adjust the graph by pressing **WINDOW** and setting **Xmin = 0, Xmax = 7, Xscl = 1, Ymin = 0** and **Ymax = .32.** Choosing Xmax = 7 leaves some space at the right of the graph in order to complete the histogram. The Ymax value was selected by looking through the values in **L2** and then rounding the largest value UP to a convenient number. Press **GRAPH** to view the histogram.

▶ Exercise 17 (pg. 212) Binomial Probabilities

A multiple-choice quiz consists of five questions and each question has four possible answers. If you randomly guess the answer to each question, the probability of getting the correct answer is 1 in 4, so p = .25 and n = 5.

a. To calculate P(X = 3), press **2ⁿᵈ [DISTR]** and select **binompdf(** and type in **5** , **.25** , **3**) and press **ENTER**.

b-c. To calculate inequalities, such as P(X ≥ 3) or P(X < 3), you can use the cumulative probability command: **binomcdf (n,p,x)**. This command accumulates probability starting at x = 0 and ending at a specified x-value. To calculate the probability of guessing at least three answers correctly, that is P(X ≥ 3), you must find P(X ≤ 2) and subtract from 1. Press **2ⁿᵈ [DISTR]** and select **binomcdf(** by scrolling through the options and selecting **binomcdf(** . Type in **5** , **.25** , **2**) and press **ENTER**. The result, P(X ≤ 2) = .896.

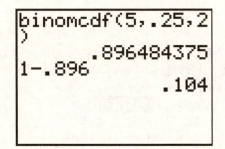

This result is the accumulated probability that X = 0, 1 or 2, since this command, **binomcdf**, accumulates probability starting at X = 0. You want P(X ≥ 3) and this is the *complement* of P(X ≤ 2). So P(X ≥ 3) = 1 - P(X ≤ 2) which is .104.
The probability of guessing *less than* 3 answers correctly is P(X ≤ 2) which equals 0.896.

◀

Section 4.3

▶ Example 1 (pg. 218) The Geometric Distribution

Based on his record, the approximate probability that LeBron James makes a free throw is 0.74. To find the probability that the *first* free throw he makes will occur on his third or fourth attempt, press **2ⁿᵈ** **[DISTR]** and select **geometpdf(** and type in **.74** , **2ⁿᵈ** **{** **3** , **4** **2ⁿᵈ** **}** **)**. This command will display 2 probabilites: P(X = 3) and P(X = 4). The probability that the first free thow he makes occurs on his third or fourth attempt is 0.050 + 0 .0130.

```
geometpdf(.74,{3
,4})
{.050024 .01300…
■
```

▶ Example 2 (pg. 219) The Poisson Distribution

Use the command **poissonpdf (μ, x)** with μ = 3 and x = 4. Press **2ⁿᵈ** **[DISTR]** and select **poissonpdf(** and type in **3** , **4** **)** and press **ENTER**. The answer will appear on the screen.

```
poissonpdf(3,4)
        .1680313557
```

▶ Exercise 16 (pg. 223)   Poisson Distribution

This is a Poisson probability problem with μ = 2.

a. To find the probability that exactly 5 businesses will file bankruptcy in any given
    minute, press **2nd [DISTR]** and select **poissonpdf(** . Type in **2 , 5)** and press
    ENTER.

b. Use the cumulative probability command **poissoncdf(.** This command accumulates
    probability starting at x = 0 and ending at the specified x-value. To calculate the
    probability that at least 5 businesses will fail in any given minute, that is P(X ≥ 5),
    you must find P(X ≤ 4) and subtract from 1. Press **2nd [DISTR]** , select **poissoncdf(**
    and type in **2 , 4)** and press **ENTER**. The result, P(X ≤ 4), is 0.947. This result
    is the accumulated probability that X = 0,1,2, 3 or 4. P(X ≥ 5) is the *complement* of
    P(X ≤ 4). So, P(X ≥ 5) is 1 − 0.947 or 0.053.

```
poissoncdf(2,4)
         .9473469827
1-.947
              .053
■
```

c. To find the probability that more than 5 businesses will fail in any given hour, find
P(X > 5). So, first calculate the *complement*, P(X ≤ 5), and then subtract the answer
from 1.

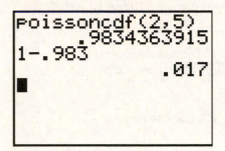

P(X > 5) = 1 − 0.983 = 0.017.

▶ Technology (pg. 233)

Exercise 1: Create a Poisson probability distribution with $\mu = 4$ for $X = 0$ to 20. Press
STAT and select **1:EDIT**. Clear **L1** and **L2**. Enter the integers 0,1,2,3,…20 into **L1** and
press **2nd** [QUIT].

For this example, it is helpful to display the probabilities with 3 decimal places. Press
MODE. Move the cursor to the 2nd line and select **3** press **ENTER** and press **2nd**
[QUIT].

Press **2nd** [DISTR] and select **B:poissonpdf(** . Type in **4** **,** **2nd** [L1] and press
ENTER.

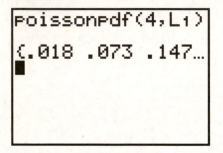

```
poissonpdf(4,L1)
{.018 .073 .147…
```

A partial display of the probabilities will appear. Press **STO** and **2nd** [L2] **ENTER**.
Press **STAT**, and select **1:EDIT. L1** and **L2** will be displayed on the screen. Notice that
the X-values in **L1** now have 3 decimal places. This is a result of setting the **MODE** to **3.**

L1	L2	L3	1
0.0000	.018	------	
1.000	.073		
2.000	.147		
3.000	.195		
4.000	.195		
5.000	.156		
6.000	.104		

L1(1)=0

Each value in **L2** is the Poisson probability associated with the X-value in **L1**. So, for
example, $P(X = 2)$ is .147. Scroll through **L2** and compare the probabilities in **L2** with
the heights of the corresponding bars in the frequency histogram on page 233.

Exercises 3 - 5: Use another technology tool to generate the random numbers. The TI-
83/84 Plus cannot generate Poisson random data.

Exercise 6: For this exercise, $\mu = 5$ and $X = 10$. Press **2nd** [DISTR] and select
B:poissonpdf(. Type in **5** **,** **10** and press **ENTER**.

Exercise 7: Use the probability distribution that you generated in Exercise 1.

a.   Sum the probabilities in **L2** that correspond to X- values of 3, 4 and 5 in **L1**.

b.   Sum the probabilities in **L2** that correspond to X- values of 0,1,2,3 and 4 in **L1**. This is $P(X \leq 4)$. Subtract this sum from **1** to get $P(X > 4)$.

c.   Assuming that the number of arrivals per minute are independent events, raise $P(X > 4)$ to the fourth power.

Normal Probability Distributions

Section 5.2

▶ Example 3 (pg. 251) Normal Probabilities

Assume triglyceride levels of the population of the United States are normally distributed with a mean of 134 and a standard deviation of 35. If you randomly select one American, calculate the probability that his/her triglyceride level is less than 80, that is P(X < 80).

The TI-83/84 Plus has two methods for calculating this probability.

Method 1: **Normalcdf**(*lowerbound*, *upperbound*, **μ, σ**) computes probability between a *lowerbound* and an *upperbound*. In this example, you are computing the probability to the left of 80, so 80 is the *upperbound*. In examples like this, where there is no *lowerbound*, you can always use *negative infinity* as the *lowerbound*. One way to indicate negative infinity is -1 x 10^99.   On the TI-83/84 Plus, use (-) 1 2nd [EE] 9 9 (Note: **EE** is the notation used on TI-83/84 Plus for raising '10' to a power. It is the second function on the comma key: ,). Try entering -1 x 10^99 by pressing (-) 1 2nd [EE] 9 9 into your calculator. The result, -1 x 10^99, is displayed as -1E99:

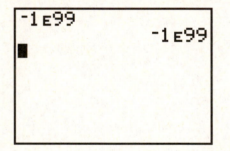

Now, to calculate P(X < 80), press **2nd** **[DISTR]** and select **2:normalcdf(** and type in **-1E99** , **80** , **134** , **35**) and press **ENTER**.

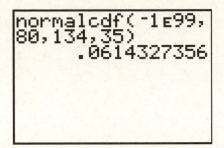

Technical note: Theoretically, the normal probability distribution (the bell-shaped curve) extends infinitely to the right (positive infinity) and to the left (negative infinity) of the mean (see Textbook pg. 236). In this particular problem, $P(X < 80)$, you do not necessarily have to use negative infinity (-1 E 99) as your *lowerbound*. If you look at this example in your textbook on pg. 251, the *lowerbound* is set at -10000. This is a perfectly fine selection since -10000 falls far below any possible minimum triglyceride level for an individual.

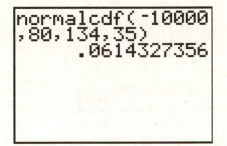

Notice that your results are the same in both of the above screens.

Method 2: This method calculates $P(X < 80)$ and also displays a graph of the probability distribution. You must first clear the Y-registers and turn **OFF** all **STATPLOTS**. Next, set up the **WINDOW** so that the graph will be displayed properly. You will need to set Xmin equal to $(\mu - 3\sigma)$ and Xmax equal to $(\mu + 3\sigma)$. Press **WINDOW** and set **Xmin** equal to $(\mu - 3\sigma)$ by entering **134 - 3 * 35.** Press **ENTER** and set **Xmax** equal to $(\mu + 3\sigma)$ by entering **134 + 3 * 35.** Set **Xscl** equal to35, the value for σ.

Setting the Y-range is a little more difficult to do. A good "rule - of - thumb" is to set **Ymax** equal to .5 /σ. For this example, type in **.5 / 35** for **Ymax.**

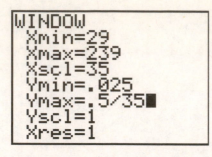

Use the Up Arrow key to highlight **Ymin**. A good value for **Ymin** is **(-) Ymax / 4** so type in [(-)] **.014 / 4**.

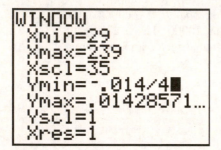

Press **2ⁿᵈ** **[QUIT]**. Clear all the previous drawings by pressing **2ⁿᵈ** **[DRAW]** (Note: **DRAW** is found above the **PRGM** key) and selecting **1:ClrDraw** and pressing **ENTER** **ENTER**. Now you can draw the probability distribution. Press **2ⁿᵈ** **[DISTR]**. Highlight **DRAW** and select **1:ShadeNorm(** and type in **-1E99** [,] **80** [,] **134** [,] **35** [)] and press **ENTER**. The output displays a normal curve with the appropriate area shaded in and its value computed.

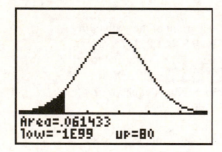

▶ Exercise 13 (pg. 253) Finding Probabilities

In this exercise, use a normal distribution with $\mu = 69.9$ and $\sigma = 3.0$.

a. Method 1: To find P(X < 66), press **2ⁿᵈ** **[DISTR]**, select **2:normalcdf(** and enter
-1E99 , 66 , 69.9 , 3) and press **ENTER**. (Note: Since no male will have a
height less than 0 inches, you could use "0" in place of –1 E 99 as your lowerbound.

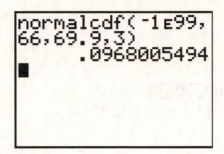

Method 2: To find P(X < 66) and include a graph, you must first clear the Y-registers and
turn **OFF** all **STATPLOTS**. Next, set up the Graph Window. Press **WINDOW** and set
Xmin = 69.9 - 3 * 3.0 and **Xmax = 69.9 + 3 * 3.0**. Set **Xscl = 3.0**. Set **Ymax = .5 / 3.0**
and Ymin = -.167/4.

Press **2ⁿᵈ** **[DRAW]** and select **1:ClrDraw** and press **ENTER** **ENTER**. Press **2ⁿᵈ**
[DISTR], highlight **DRAW** and select **1:ShadeNorm(** and type in
-1E99 , 66 , 69.9 , 3) and press **ENTER**.

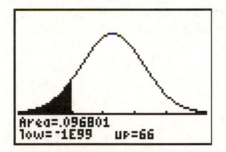

b. To find P(66< X < 72), press **2ⁿᵈ** **[DISTR]**, select **2:normalcdf(** and type in **66** , **72**
 , **69.9** , **3**) and press **ENTER**.

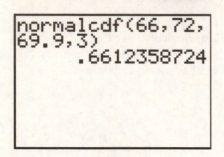

Or, press **2ⁿᵈ** **[DRAW]** and select **1:ClrDraw** and press **ENTER** **ENTER**. Press **2ⁿᵈ**
[DISTR] , highlight **DRAW** and select **1:ShadeNorm(** and type in **66** , **72** , **69.9** ,
3) and press **ENTER**.

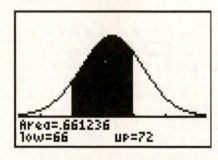

c. To find P(X > 72), press **2ⁿᵈ** **[DISTR]**, select **2:normalcdf(** and type in **72** , **1E99** ,
 69.9 , **3**) and press **ENTER**. (Note: In this example, the lowerbound is 72 and the
 upperbound is positive infinity).

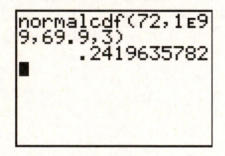

Or, press **2ⁿᵈ** **[DRAW]** and select **1:ClrDraw** and press **ENTER** **ENTER**. Press **2ⁿᵈ**
[DISTR] , highlight **DRAW** and select **1:ShadeNorm(**and type in **72** , **1E99** , **69.9** ,
3) and press **ENTER**.

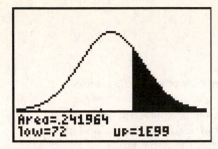

Note: When using the TI-83/84 Plus (or any other technology tool), the answers you obtain may vary slightly from the answers that you would obtain using the standard normal table. Consequently, your answers may not be exactly the same as the answers found in the answer key in your textbook. The differences are simply due to rounding.

Section 5.3

▶ Example 4 (pg. 260) Finding a Specific Data Value

This is called an inverse normal problem and the command **invNorm(area, μ,σ)** is used. In this type of problem, a percentage of the area under the normal curve is given and you are asked to find the corresponding X-value. In this example, the percentage given is the top 10%. The TI-83/84 Plus always calculates probability from negative infinity up to the specified X-value. To find the X-value corresponding to the top 10%, you must accumulate the bottom 90% of the area. Press **2ⁿᵈ** **[DISTR]** and select **3:invNorm(** and type in **.90** **,** **50** **,** **10** **)** and press **ENTER**.

```
invNorm(.90,50,1
0)
        62.81551567
```

In order to score in the top 10%, you must earn a score of at least 62.82. Assuming that scores are given as whole numbers, your score must be at least 63.

◀

▶ Exercise 32 (pg. 263) Heights of Males

This is a normal distribution with μ = 69.9 and σ = 3.

a. To find the 90th percentile, press **2ⁿᵈ [DISTR]** and select **3:invNorm(** and type in **.90** **,** **69.9** **,** **3** **)** and press **ENTER**.

```
invNorm(.90,69.9
,3)
        73.7446547
```

b. To find the first quartile, press **2ⁿᵈ [DISTR]** and select **3:invNorm(** and type in **.25** **,** **69.9** **,** **3** **)** and press **ENTER**.

```
invNorm(.25,69.9
,3)
        67.87653075
```

◀

Section 5.4

▶ Example 4 (pg. 271) Probabilities for Sampling Distributions

In this example, data has been collected on the average daily driving time for different age groups. From the graph on pg. 271, you will find that the mean driving time for adults in the 15 to 19 age group is: $\mu = 25$ minutes. The problem states that the assumed standard deviation is $\sigma = 1.5$ minutes.

You randomly sample 50 drivers in the 15 – 19 age group. Since the sample size, n, is greater than 30, you can conclude that the sampling distribution of the sample mean is approximately normal with $u_{\bar{x}} = 25$ and $\sigma_{\bar{x}} = 1.5/\sqrt{(50)}$. To calculate P($24.7 < \bar{x} <$ 25.5), press **2ⁿᵈ** **[DISTR]**, select **2:normalcdf(** and type in **24.7** **,** **25.5** **,** **25** **,** **1.5/√(50))** **)** and press **ENTER**.

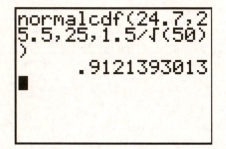

```
normalcdf(24.7,2
5.5,25,1.5/√(50)
)
          .9121393013
■
```

Note: The answer in your textbook is 0.9116. This answer was calculated using the z-table. Since z-values in the table are rounded to hundredths, the answers will vary slightly from those obtained using the TI-83/84 Plus.

◀

▶ Example 6 (pg. 273) Finding Probabilities for x and $\bar{x}$

The population is normally distributed with μ= 3173 and σ = 1120.

1. To calculate P(X < 2700), press **2nd** **[DISTR]**, select **2:normalcdf(** and type in **-1E99** ⟨,⟩ **2700** ⟨,⟩ **3173** ⟨,⟩ **1120** ⟨)⟩ and press **ENTER**. (Note: Since the minimum credit card balance is 0, the lowerbound could be set at 0, rather than negative infinity.)

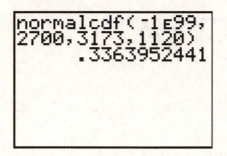

2. To calculate P($\bar{x}$ < 2700), press **2nd** **[DISTR]**, select **2:normalcdf(** and type in **-1E99** ⟨,⟩ **2700** ⟨,⟩ **3173** ⟨,⟩ **1120/√(25)** ⟨)⟩ ⟨)⟩ and press **ENTER**.

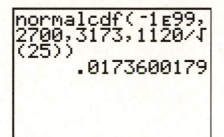

◀

▸ Exercise 33 (pg. 277) Make a Decision

To decide whether the machine needs to be reset, you must decide how unlikely it would be to find a mean of 127.9 from a sample of 40 cans if, in fact, the machine is actually operating correctly at $\mu = 128$. One method of determining the likelihood of $\bar{x} = 127.9$ is to calculate how far 127.9 is from the mean of 128. You can do this by calculating how much area there is under the normal curve to the left of 127.9. The smaller that area is, the farther 127.9 is from the mean and the more unlikely 127.9 is.

To calculate P($\bar{x} <$ 127.9), press **2**nd **[DISTR]** , select **2:normalcdf(** and type in **-1E99** , **127.9** , **128** , **0.20/√(40)**)) and press **ENTER**.

```
normalcdf(-1E99,
127.9,128,.20/√(
40))
    7.827669991E-4
```

Notice that the answer is displayed in scientific notation: 7.827E-4. Convert this to standard notation, .0007827, by moving the decimal point 4 places to the left. This probability is extremely small; therefore, the event ($\bar{x} <$ 127.9) is highly unlikely if the mean is actually 128. So, something has gone wrong with the machine and the actual mean must have shifted to a value less than 128.

◂

▶ Technology (pg. 299) Age Distribution in the U. S.

Exercise 1: Press **STAT** and select **1:EDIT**. Clear **L1**, **L2** and **L3**.
Enter the age distribution into **L1** and **L2** by putting class midpoints in **L1** and relative
frequencies (converted to decimals) into **L2**. (The first entry is **2** in **L1** and **.069** in **L2**.)
To find the population mean, μ, and the population standard deviation, σ, press **STAT**,
highlight **CALC**, select **1:1-Var Stats**, press **ENTER** and press **2**nd **[L1]** **,** **2**nd **[L2]**
ENTER. The mean and the standard deviation will be displayed. (Note: Use " σ x"
because the age data represents the entire population distribution of ages, not a sample.)

Exercise 2: Enter the thirty-six sample means into **L3**. To find the mean and standard
deviation of these sample means, press **STAT**, highlight **CALC**, select **1:1-Var Stats**.
Press **ENTER** and press **2**nd **[L3]** **ENTER**. The mean and the standard deviation will be
displayed. (Note: use "sx" because the 36 sample means are a sample of 36 means, not the
entire population of all possible means of size n = 40).

Exercise 3: Use the histogram on pg. 299 to answer this question.

Exercise 4: The TI-83/84 Plus will draw a frequency histogram for a set of data, not a
relative frequency histogram. (The shape of the data can be determined from either type
of histogram). Press **2**nd **[STAT PLOT]** and select **1: Plot 1.** Turn **ON** Plot 1, select
Histogram for **Type,** set **Xlist** to **L3** and set **Freq** to **1**. Press **ZOOM** and **9** for
ZoomStat. To adjust the histogram so that it has nine classes, press **WINDOW**. For
Xmin, use a value slightly smaller than the minimum value of the sample means; for
Xmax, use a value slightly larger than the maximum value of the sample means;
approximate the class width by finding the range of values (max – min) of the sample
means and dividing this range by 9. This value is approximately 2, so set **Xscl = 2** and
press **GRAPH**.

Exercise 5: See the output from Exercise 1 for the population standard deviation, σ.

Exercise 6: See the output from Exercise 2 for the standard deviation of the 40 sample
means. This standard deviation, $s_{\bar{x}}$, is an approximation of $\sigma_{\bar{x}}$. The Central Limit

Theorem states that $\sigma_{\bar{x}} \approx \dfrac{\sigma}{\sqrt{n}}$. Use your results to confirm this fact.

Section 6.1

▶ Example 4 (pg. 308) Constructing a Confidence Interval for μ (Large Sample Case)

Enter the data from Example 1 on pg. 304 into **L1**. The sample standard deviation, s, for this data set is approximately 53.0. In this example, n > 30, so the sample standard deviation is a good approximation of σ, the population standard deviation. Using this sample standard deviation as an estimate of σ, you can construct a Z-Interval, a confidence interval for μ, the population mean. Press **STAT**, highlight **TESTS** and select **7:Zinterval.**

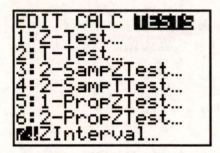

On the first line of the display, you have the option to select **Data** or **Stats.** For this example, select **Data** because you are using the actual data which is in **L1**. Press **ENTER**. Move to the next line and enter 53.0, the estimate of **σ.** On the next line, enter **L1** for **LIST**. For **Freq**, enter **1**. For **C-Level**, enter **.99** for a 99% confidence interval. Move the cursor to **Calculate.**

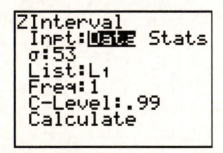

Press **ENTER**.

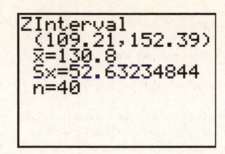

A 99% confidence interval estimate for μ, the population mean is (109.21, 152.39). The output display includes the sample mean (130.8), the sample standard deviation (52.63), and the sample size (40).

◄

▶ Example 5 (pg. 309) Confidence Interval for μ (σ known)

In this example, $\bar{x} = 22.9$ years, n = 20, a value for σ from previous studies is given as σ = 1.5 years, and the population is normally distributed. Construct a 90% confidence interval for μ, the mean age of all students currently enrolled at the college.

Press **STAT**, highlight **TESTS** and select **7:ZInterval**. In this example, you do not have the actual data. Instead you have the summary statistics of the data, so select **Stats** and press **ENTER**. Enter the value for **σ: 1.5**; enter the value for $\bar{x}$ **: 22.9**; enter the value for **n: 20**; and enter **.90** for **C-level**. Highlight **Calculate**.

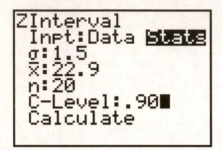

Press **ENTER**.

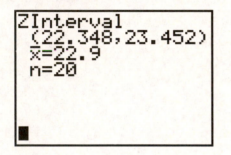

Using a 90% confidence interval, you estimate that the average age of all students is between 22.3 and 23.5 years.

◀

▶ Exercise 49 (pg. 313) Confidence Interval for μ, Large Sample Case

Enter the data into **L1**. Press **STAT**, highlight **TESTS** and select **7:ZInterval**. Since you have entered the actual data points into **L1**, select **Data** for **Inpt** and press **ENTER**. Enter the value for σ, **1.3**. Set **LIST = L1, Freq = 1** and **C-level = .90**. Highlight **Calculate.**

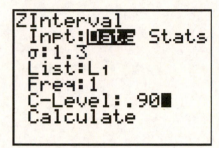

Now press **ENTER**.

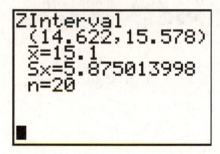

Repeat the process and set **C-level = .99**.

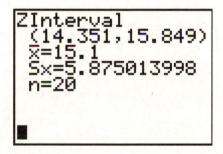

▶ Exercise 63 (pg. 315) Using Technology

Notice in the stemplot for this example that there are several repetitive values (i.e., the value of 213 appears 5 times). Due to these repeats it may be more efficient to enter the data in table form. Begin by entering all the unique values into L1. Then enter the frequencies into L2.

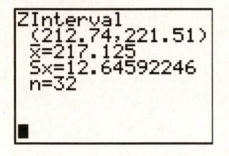

In this example, no estimate of σ is given so you should use s, the sample standard deviation as a good approximation of σ since n > 30. Press **STAT**, highlight **CALC**, select **1:1-Var Stats** and press **2ⁿᵈ** **[L1]** **,** **2ⁿᵈ** **[L2]** **ENTER**. The sample standard deviation, Sx, is 12.65.

Press **STAT**, highlight **TESTS** and select **7:ZInterval**. For **Inpt**, select **Data**. For σ, enter **12.65**. Set **List** to **L1**, **Freq** to **L2** and **C-level** to **.95**. Highlight **Calculate** and press **ENTER**.

```
ZInterval
 (212.74,221.51)
 x̄=217.125
 Sx=12.64592246
 n=32
```

Section 6.2

▸ Example 3 (pg.321)   Constructing a Confidence Interval for μ (Small Sample Case)

In this example, n = 20, $\bar{x}$ = 9.75, s = 2.39 and the underlying population is assumed to be normal. Find a 99% confidence interval for μ. First, notice that σ is unknown. The sample standard deviation, s, is not a good approximation of σ when n < 30. To construct the confidence interval for μ, the correct procedure under these circumstances (n < 30, σ is unknown and the population is assumed to be normally distributed) is to use a T-Interval.

Press **STAT**, highlight **TESTS**, scroll through the options and select **8:TInterval** and press **ENTER**. Select **Stats** for **Inpt** and press **ENTER**. Fill in $\bar{x}$, **Sx**, and **n** with the sample statistics. Set **C-level** to **.99**. Highlight **Calculate**.

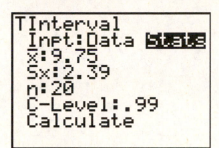

Press **ENTER**.

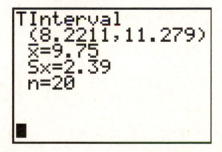

► Exercise 27 (pg. 324) Deciding on a Distribution

In this example, n = 50, $\bar{x}$ = 27.7 and s = 6.12. Notice that σ is unknown. You can use s, the sample standard deviation, as a good approximation to σ in this case because n > 30. To calculate a 95% confidence interval for μ, press **STAT**, highlight **TESTS** and select **7:ZInterval**. Fill in the screen with the appropriate information.

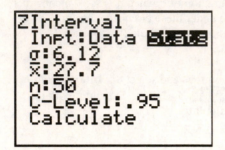

Calculate the confidence interval.

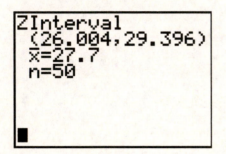

▶ Exercise 29 (pg. 325) Deciding on a Distribution

Enter the data into **L1**. In this case, with n = 25 and σ unknown, and assuming that the population is normally distributed, you should use a Tinterval. Press **STAT**, highlight **TESTS** and select **8:TInterval**. Fill in the appropriate information.

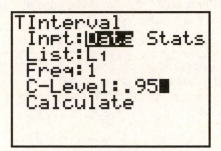

Calculate the confidence interval.

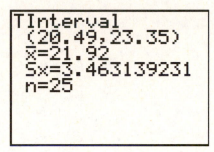

Section 6.3

▶ Example 2 (pg. 329)   Constructing a Confidence Interval for p

In this example, 1000 U.S. adults are surveyed and 662 say that it is acceptable to check personal e-mail at work (see Example 1, pg. 327). Construct a 95% confidence interval for p, the proportion of all Americans who say that it is acceptable to check personal e-mail at work.

Press **STAT**, highlight **TESTS**, scroll through the options and select **A:1-PropZInt**. The value for X is the number of U.S. adults in the group of 1000 who said that it is acceptable to check personal e-mail at work, so **X = 662**. The number who were surveyed is n, so **n = 1000**. Enter **.95** for **C-level**.

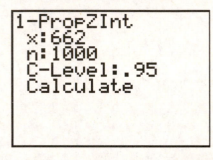

Highlight **Calculate** and press **ENTER**.

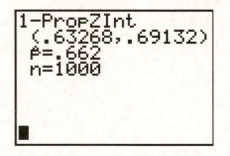

In the output display the confidence interval for p is (.63268, .69132). The sample proportion, $\hat{p}$, is .662 and the number surveyed is 1000.

▶ Example 3 (pg. 330) Confidence Interval for p

From the graph, the sample proportion, $\hat{p}$, is 0.71 and n is 498. Construct a 99% confidence interval for the proportion of adults who think that teenagers are the more dangerous drivers. In order to construct this interval using the TI-83/84 Plus, you must have a value for X, the number of people in the study who said that teenagers were the more dangerous drivers. Multiply 0.71 by 498 to get this value. If this value is not a whole number, round to the nearest whole number. For this example, X is 354.

Press **STAT**, highlight **TESTS** and select **A:1-PropZInt** by scrolling through the options or by simply pressing **ALPHA** **A**. Enter the appropriate information from the sample.

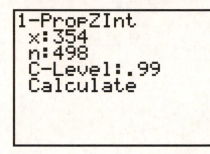

Highlight **Calculate** and press **ENTER**.

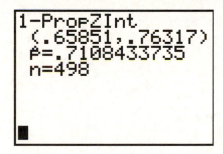

▶ Exercise 21 (pg. 334) Confidence Interval for p

a.   To construct a 99% confidence interval for the proportion of adults from the United
     States who believe that climate change poses a large threat to the world, begin by
     multiplying .27 by 1017 to obtain a value for X, the number of adults in the U.S.
     sample who said that climate change poses a large threat to the world. Press **STAT**,
     highlight **TESTS** and select **A:1-PropZInt**. Fill in the appropriate values.

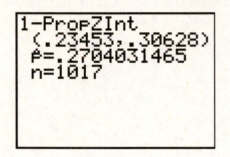

```
1-PropZInt
 x:275
 n:1017
 C-Level:.99■
 Calculate
```

Highlight **Calculate** and press **ENTER**.

```
1-PropZInt
 (.23453,.30628)
 p̂=.2704031465
 n=1017
```

b. and c. Follow the same procedure as shown in part a.

> ▶ Technology (pg. 351)   "Most Admired" Polls

1. Use the survey information to construct a 95% confidence interval for the proportion of the population who would have chosen Barack Obama as their most admired man. Multiply .30 by 1025 to obtain a value for X, the number of people in the sample who chose Barack Obama as their most admired man. Press **STAT**, highlight **TESTS** and select **A:1-PropInt**. Enter the values for **X** and **n**. Set **C-level** to **.95**, highlight **Calculate** and press **ENTER**.

2. To construct a 95% confidence interval for the proportion of the population who would have chosen Hillary Clinton as their most admired female, you must calculate X, the number of people in the sample who chose Hillary Clinton. Multiply .16 by 1025 and round your answer to the nearest whole number. Press **STAT**, highlight **TESTS**, select **A:1-PropZInt** and fill in the appropriate information. Press **Calculate** and press **ENTER**.

4. To construct a 95% confidence interval for the proportion of the population who would have chosen Sarah Palin as their most admired female, you must calculate X, the number of people in the sample who chose Sarah Palin. Multiply .15 by 1025 and round your answer to the nearest whole number. Press **STAT**, highlight **TESTS**, select **A:1-PropZInt** and fill in the appropriate information. Press **Calculate** and press **ENTER**.

5. To do one simulation, press **MATH**, highlight **PRB**, select **7:randBin(** and type in **1025** $\boxed{,}$ **.18** $\boxed{)}$. The output is the number of successes in a survey of n = 1025 people. In this case, a "success" is choosing Sarah Palin as your most admired female. (Note: It takes approximately one minute for the TI-83/84 Plus to do the calculation).

    To run the simulation ten times use **7:randBin(1025, .18, 10)**. (These simulations will take approximately 8 minutes). The output is a list of the number of successes in each of the 10 simulations. Calculate $\hat{p}$ for each of the simulations. $\hat{p}$ is equal to: (number of successes)/1025. Use these 10 values of $\hat{p}$ to answer questions 5a and 5b.

◀

Hypothesis Testing with One Sample

Section 7.2

▸ Example 4 (pg. 374) Hypothesis Testing Using P-values

The hypothesis test, H$_o$: $\mu \geq 13$ vs. H$_a$: $\mu < 13$, is a left-tailed test. The sample statistics are $\bar{x} = 12.9$, s = 0.19 and n = 32. The sample size is greater than 30, so the **Z-Test** is the appropriate test. To run the test, press **STAT**, highlight **TESTS** and select **1:Z-Test**. Since you are using sample statistics for the analysis, select **Stats** for **Inpt** and press **ENTER**. For μ_o enter 13, the value for μ in the null hypothesis. For σ enter **s**, the sample standard deviation. (Note: s, the sample standard deviation, is a good approximation for σ, the population standard deviation, when n is large.) Enter **12.9** for $\bar{x}$ and **32** for **n**. On the next line, choose the appropriate alternative hypothesis and press **ENTER**. For this example, it is < μ_o, a left-tailed test.

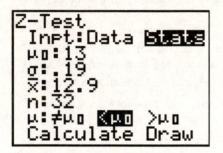

There are two choices for the output of this test.

The first choice is **Calculate**. The output displays the alternative hypothesis, the calculated z-value, the P-value, $\bar{x}$, and n.

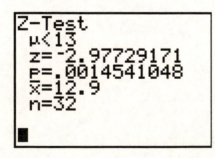

To view the second output option, you should start by turning **Off** any **STATPLOT** that is turned **ON**. Press **2nd** **Y=** to display the **STATPLOTS**. If a **PLOT** is **On**, select it and move the cursor to Off and press **ENTER**. Also, clear all Y-registers.

Press **STAT**, highlight **TESTS**, and select **1:Z-Test**. All the necessary information for this example is still stored in the calculator. Scroll down to the bottom line and select **DRAW**. A normal curve is displayed with the left-tail area of .0015 shaded. This shaded area is the area to the left of the calculated Z-value. (Because the area is so small in this example, it is not actually visible in the diagram.) The Z-value and the P-value are also displayed.

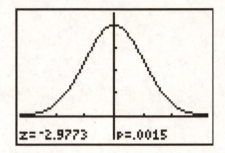

Since p = .0015, which is less than α, the correct conclusion is to **Reject H₀.** There is enough evidence at the 1% level of significance to support the claim that the mean pit stop time is less than 13 minutes.

◄

▶ Example 5 (pg. 375)   Hypothesis Testing Using P-values

This test is a two-tailed test for H₀: μ = $22,500 vs. Hₐ: μ ≠ $22,500. The sample
statistics are $\bar{x}$ = $21,545, s = $3,015 and n = 30. Press **STAT**, highlight **TESTS** and
select **1:Z-Test**. Choose **Stats** for **Inpt** and press **ENTER**. For μ₀ enter 22500, the value
for μ in the null hypothesis. For **σ**, enter **s**, the sample standard deviation. Enter **21545**
for $\bar{x}$ and **30** for **n**. On the next line, choose the appropriate alternative hypothesis and
press **ENTER**. For this example, it is ≠ μ₀, a two-tailed test.

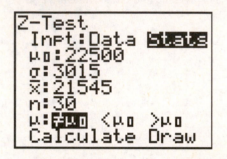

Highlight **Calculate** and press **ENTER**.

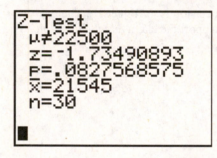

Or, highlight **Draw** and press **ENTER**.

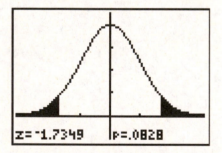

Notice the P-value is equal to .0828. In this example, α is .05. Since the P-value is
greater than α, the correct conclusion is to **Fail to Reject H₀**. There is not enough
evidence at the 5% level of significance to support the claim that the mean cost of
bariatric surgery is different from $22,500.

◀

▸ Exercise 29 (pg. 383) Testing Claims Using P-values

Test the hypotheses: H_o: $\mu \leq 30$ vs. H_a: $\mu > 30$. Since n > 30, the appropriate test is the **Z-Test**. This procedure requires a value for σ. In cases with n > 30, the sample standard deviation, s, can be used to approximate σ. The sample statistics are $\bar{x} = 31$, s = 2.5 and n = 50. Press **STAT**, highlight **TESTS** and select **1:Z-Test**. For **Inpt**, choose **Stats** and press **ENTER**. Fill the input screen with the appropriate information. Choose $> \mu_o$ for the alternative hypothesis and press **ENTER**.

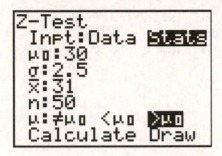

Highlight **Calculate** and press **ENTER**.

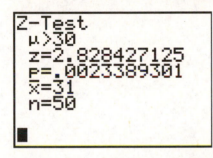

Or, highlight **Draw** and press **ENTER**.

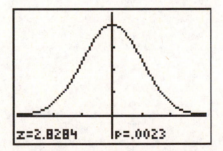

The test statistic is Z = 2.8284 and the P-value is .0023. Since the P-value is less than α, the correct conclusion is to **Reject H_o**. There is enough evidence at the 1% level of significance to support the student's claim that the mean raw score for the school's applicants is more than 30.

◀

▶ Exercise 33 (pg. 383) Testing Claims Using P-values

This is a hypothesis test for H_o: $\mu = 15$ vs. H_a: $\mu \neq 15$. Since n > 30, the appropriate test
is the **Z-Test**. This procedure requires a value for σ. In cases with n > 30, the sample
standard deviation, s, can be used to approximate σ. Begin the analysis by entering the
32 data points into **L1**. Press **STAT**, highlight **CALC** and choose **1:1-Var Stats** and
press 2nd **L1** **ENTER**. The sample statistics will be displayed on the screen. The value
you need for the hypothesis test is Sx, the sample standard deviation, which equals 4.29.

To perform the test, press **STAT**, highlight **TESTS** and select **1:Z-Test**. Since you have
the actual data points for the analysis, select **Data** for **Inpt** and press **ENTER**. Fill in the
input screen with the appropriate information and select $\neq \mu_o$ as the alternative hypothesis
and press **ENTER**.

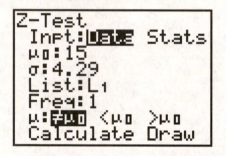

Highlight **Calculate** and press **ENTER**

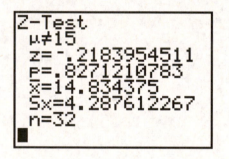

Or, highlight **Draw** and press **ENTER**.

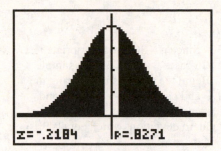

Since the P-value is greater than α, the correct conclusion is to **Fail to Reject H₀**. There is not enough evidence at the 5% level of significance to reject the claim that the mean amount of time it takes smokers to permanently quit smoking is 15 years.

◀

▸ Exercise 39 (pg. 384) Testing Claims

This is a hypothesis test for $H_o: \mu \le 32$ vs. $H_a: \mu > 32$. Since n > 30, the appropriate test is the **Z-Test**. This procedure requires a value for σ. In cases with n > 30, the sample standard deviation, s, can be used to approximate σ. Begin the analysis by entering the 34 data points into **L1**. Press **STAT**, highlight **CALC** and choose **1:1-Var Stats** and press **2ⁿᵈ L1 ENTER**. The sample statistics will be displayed on the screen. The value you need for the hypothesis test is Sx, the sample standard deviation, which equals 9.16.

To run the hypothesis test, press **STAT**, highlight **TESTS** and select **1:Z-Test**. Since you have the actual data points for the analysis, select **Data** for **Inpt** and press **ENTER**. Enter **32** for μ_o and enter **9.16** for σ. The data is stored in **L1** and **Freq** is **1**. The alternate hypothesis is a right-tailed test so select $> \mu_o$ and press **ENTER**.

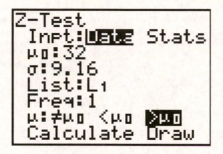

To run the test, select **Calculate** and press **ENTER**.

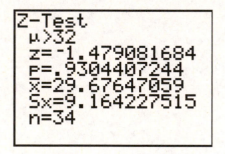

Or, select **Draw** and press **ENTER**.

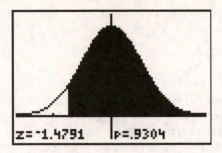

z=-1.4791   P=.9304

Since the P-value of .9304 is greater than α, the correct conclusion is to **Fail to Reject H₀.** There is not enough evidence at the 6% level of significance to support the scientist's claim that the mean nitrogen dioxide level in Calgary is greater than 32 parts per billion.

Section 7.3

▸ Example 6 (pg. 393)   Using P-values with a T-Test

This test is a two-tailed test of H_o: $\mu \geq 14$ vs. H_a: $\mu < 14$. The sample statistics are $\bar{x} = 13$, $s = 3.5$ and $n = 10$. Since $n < 30$, the test is a **T-Test** if you assume that the underlying population is approximately normally distributed. Press **STAT**, highlight **TESTS** and select **2:T-Test**. Choose **Stats** for **Inpt** and press **ENTER**. Fill in the following information: $\boldsymbol{\mu_o = 14}$, $\boldsymbol{\bar{x} = 13}$, **Sx = 3.5** and **n = 10**. Choose the one-tailed alternative hypothesis, **< μ_o**, and press **ENTER**.

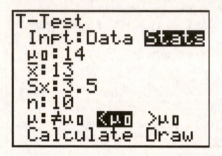

Highlight **Calculate** and press **ENTER**

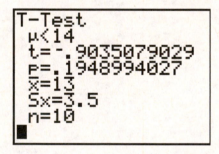

Or, highlight **Draw** and press **ENTER**.

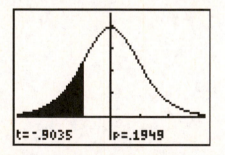

Since the P-value is greater than α, the correct conclusion is to **Fail to Reject H_o**. There is not enough evidence at the 10% level of significance to support the office's claim that the mean wait time is less than 14 minutes.

◂

▶ Exercise 29 (pg. 396) Testing Claims Using P-values

The correct hypothesis test is H_o: $\mu \geq 32$ vs. H_a: $\mu < 32$. Enter the data into L1. Since n < 30, the appropriate test is the **T-Test**, if you assume that the underlying population is approximately normal. Press **STAT**, highlight **TESTS** and select **2:T-Test**. Select **Data** for **Inpt** and press **ENTER**. Fill in the screen with the necessary information.

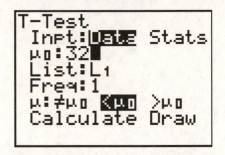

Choose **Calculate** and press **ENTER**.

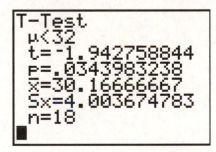

Or, choose **Draw** and press **ENTER**.

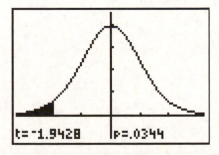

Since the P-value is less than α, the correct conclusion is to **Reject H_o**. There is enough evidence at the 5% level of significance to support the claim that the mean class size for full-time faculty is less than 32 students.

◀

▶ Exercise 37 (pg. 397) Deciding on a Distribution

This test is a left-tailed test of H_o: $\mu \geq 23$ vs. H_a: $\mu < 23$. The sample statistics are $\bar{x} = 22$, $s = 4$ and $n = 5$. Since $n < 30$ and gas mileage is normally distributed, the appropriate test is the **T-Test**. Press **STAT**, highlight **TESTS** and select **2:T-Test**. Fill in the screen with the necessary information.

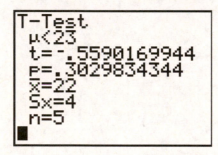

Choose **Calculate** and press **ENTER**.

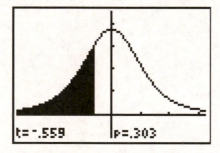

Or, choose **Draw** and press **ENTER**.

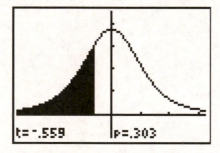

Since the P-value is greater than α, the correct conclusion is to **Fail to Reject H_o.** There is not enough evidence at the 5% level of significance to reject the car company's claim that the mean gas mileage for the luxury sedan is at least 23 miles per gallon.

◀

▸ Exercise 38 (pg. 397) Deciding on a Distribution

This test is a right-tailed test of H_o: $\mu \le \$35,000$ vs. H_a: $\mu > \$35,000$. The sample statistics are $\bar{x} = \$34,967$, $s = \$5933$ and $n = 50$. Since $n \ge 30$, s is a good approximation of σ, so the appropriate test is the **Z-Test**. Press **STAT**, highlight **TESTS** and select **1:Z-Test**. Fill in the screen with the necessary information.

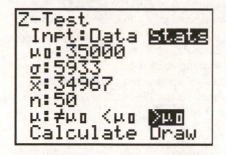

Choose **Calculate** and press **ENTER**.

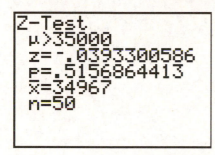

Or, choose **Draw** and press **ENTER**.

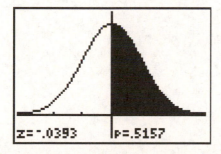

Since the P-value is greater than α, the correct conclusion is to **Fail to Reject H_o.** There is not enough evidence at the 1% level of significance to support the education publication's claim that the average in-state tuition for one year of law school is more than $35,000.

◂

Section 7.4

▶ Example 1 (pg. 399) Hypothesis Test for a Proportion

This hypothesis test is a left-tailed test of: H_o: $p \geq .50$ vs. H_a: $p < .50$. The first step is to verify that the products 'np' and 'nq' are both great than 5. Since both products are, in fact, greater than 5, the **1-Proportion Z-test** is the appropriate test. Press **STAT**, highlight **TESTS** and select **5:1-PropZTest**. This test requires a value for p_o, which is the value for **p** in the null hypothesis. Enter **.50** for p_o. Next, a value for X is required. X is the number of "successes" in the sample. In this example, a success is "having accessed the Internet over a wireless network with a laptop computer." Since 39% of the 100 individuals in the sample say that they have accessed the Internet over a wireless network with a laptop computer, **X** is equal to .39 times 100 or **39**. Next, enter the value for n. Select **< p_o** for the alternative hypothesis and press **ENTER**.

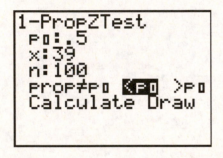

Highlight **Calculate** and press **ENTER**.

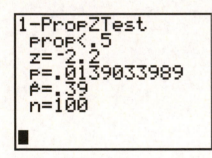

Or, highlight **Draw** and press **ENTER**.

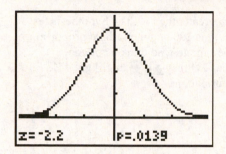

z=-2.2   p=.0139

Since the P-value is greater than α, the correct conclusion is to **Fail to Reject H₀**. There is not enough evidence at the 1% level of significance to support the claim that less than 50% of U.S. adults have accessed the Internet over a wireless network with a laptop computer

◀

▶ Example 2 (pg. 400) Hypothesis Test for a Proportion

First, verify that the products 'np' and 'nq' are both greater than 5. The hypothesis test is: H_o: p = 0.25 vs. H_a: p ≠ 0.25. The sample size, n, is 200 and the sample proportion, $\hat{p}$, is 0.21. **X**, the number of college graduates in the sample who think a college education is not worth the cost, is 200 times 0.21 or **42**. Press **STAT**, highlight **TESTS** and select **5:1-PropZTest**. Enter the necessary information.

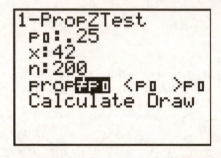

Highlight **Calculate** and press **ENTER**.

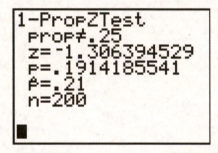

Or, highlight **Draw** and press **ENTER**.

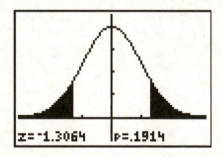

Since the P-value is greater than α, the correct conclusion is to **Fail to Reject H_o**. There is not enough evidence at the 10% level of significance to reject the claim that 25% of college graduates think a college degree is not worth the cost.

                                                                                    ◀

▶ Exercise 11 (pg. 401) Testing Claims

Use a right-tailed hypothesis test to test the hypotheses: H_o: $p \leq 0.50$ vs. H_a: $p > 0.50$. The sample statistics are n = 150 and $\hat{p}$ = 0.58. Multiply n times $\hat{p}$ to find X, the number of people in the sample who believe that drivers should be allowed to use cellular phones with hands-free devices while driving. Press **STAT**, highlight **TESTS** and select **5:1-PropZTest**. Enter the necessary information.

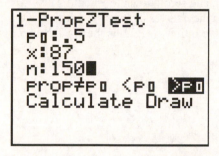

Highlight **Calculate** and press **ENTER**.

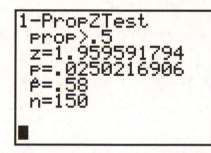

Or, highlight **Draw** and press **ENTER**.

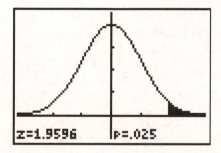

Since the P-value is greater than α, the correct conclusion is to **Fail to Reject H_o**. There is not enough evidence at the 1% level of significance to reject the claim that at most 50% of the population believe that drivers should be allowed to use cellular phones with hands-free devices while driving.

◀

Section 7.5

▶ Example 4 (pg. 407)   Hypothesis Test for a Variance

This hypothesis test is a right-tailed test of: H_o: $\sigma^2 \leq 0.25$ vs. H_a:$\sigma^2 > 0.25$. The sample statistics are $s^2 = 0.27$ and n = 41. First you must calculate the χ^2 value: $((41-1)$ $*0.27)/0.25 = 43.2$. Next, find the P-value associated with this χ^2 value. Press 2nd **DISTR**, (**DISTR** is the 2nd function on **VARS** key), and scroll down to χ^2 cdf(. This function requires a *lowerbound*, an *upperbound* and the *degrees of freedom*. Since this is a right-tailed test, 43.2 is the *lowerbound* and positive infinity (1E99) is the *upperbound*. The degrees of freedom (n-1) is equal to 40.

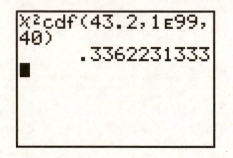

The P-value is 0.336. Since this is greater than α, the correct conclusion is to **Fail to Reject H_o**. There is not enough evidence at the 5% level of significance to reject the company's claim that the variance of the amount of fat in the whole milk is not more than 0.25.

◀

▶ Technology (pg. 423) The Case of the Vanishing Women

Exercise 1: Use the TI-83/84 Plus to run the hypothesis test and compare your results to the MINITAB results shown in the display at the bottom of pg. 423. Use a two-tailed test to test the hypotheses: H_o: p = 0.53 vs. H_a: p≠ 0.53. The sample statistics are X = 102 women and n = 350 people selected from the Boston City Directory. Press **STAT**, highlight **TESTS** and select **5:1-PropZTest**. Fill in the appropriate information highlight **Calculate** and press **ENTER** or, highlight **Draw** and press **ENTER**.

Exercise 4: In the first stage of the jury selection process, 350 people are selected and 102 of them are women. So, at this stage, the proportion of women is 102 out of 350, or 0.2914. From this population of 350 people, a sample of 100 people is selected at random and only nine are women. Test the claim that the proportion of women in this population of 350 people is 0.2914. Use a two-tailed test to test the hypotheses: H_o: p = 0.2914 vs. H_a: p≠ 0.2914. The sample statistics are X = 9 women and n = 100 people. Press **STAT**, highlight **TESTS** and select **5:1-PropZTest**. Fill in the appropriate information, Highlight **Calculate** and press **ENTER** or highlight **Draw** and press **ENTER**.

◀

CHAPTER 8

Hypothesis Testing with Two Samples

Section 8.1

▸ Example 3 (pg. 433) Two-Sample Z-Test

Test the claim that the average daily cost for meals and lodging when vacationing in Texas is less than the average costs when vacationing in Virginia. Designate Texas as population 1 and Virginia as population 2. The appropriate hypothesis test is a left-tailed test of $H_o: \mu_1 \geq \mu_2$ vs. $H_a: \mu_1 < \mu_2$. The sample statistics are displayed in the table at the top of pg. 433. Each sample size is greater than 30, so the correct test is a Two-sample Z-Test.

Press **STAT**, highlight **TESTS** and select **3:2-SampZTest**. Since you are using the sample statistics for your analysis, select **Stats** for **Inpt** and press **ENTER**. In order to use this test, values for σ_1 and σ_2 are required. The sample standard deviations, s_1 and s_2, can be used as approximations for σ_1 and σ_2 when both n_1 and n_2 are greater than or equal to 30. Enter 18 for σ_1. Enter 24 for σ_2. Next, enter the mean and sample size for Group1: $\bar{x}_1 = 216$ and $n_1 = 50$. Continue by entering the mean and sample size for Group 2: $\bar{x}_2 = 222$ and $n_2 = 35$.

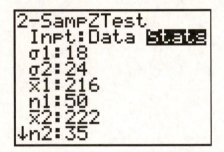

Use the down arrow to display the next line. This line displays the three possible alternative hypotheses for testing: $\mu_1 \neq \mu_2$, $\mu_1 > \mu_2$, or $\mu_1 < \mu_2$. For this example, select $< \mu_2$ and press **ENTER**. Scroll down to the next line and select **Calculate** or **Draw**.

The output for **Calculate** is displayed on two pages. (Notice that the only piece of information on the second page is n_2, so page 2 is not displayed here.)

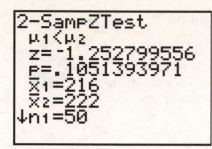

The output for **Draw** contains a graph of the normal curve with the area associated with the test statistic shaded.

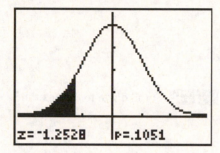

Both outputs display the P-value, which is .1051. Since the P-value is greater than α, the correct decision is to **Fail to Reject H₀.** At the 1% level of significance, there is not enough evidence to support the claim that the average cost for meals and lodging when vacationing in Texas is less than the average cost when vacationing in Virginia.

◀

▶ Exercise 21 (pg. 436) Testing the Difference between Two Population Means

The appropriate hypothesis test is a two-tailed test of $H_o: \mu_1 = \mu_2$ vs.. $H_a: \mu_1 \neq \mu_2$. Designate Tire Type A as population 1 and Tire Type B as population 2. Press **STAT**, highlight **TESTS** and select **3:2-SampZTest**. For **Inpt,** select **Stats** and press **ENTER**. Enter the sample standard deviations as approximations of σ_1 and σ_2. Next, enter the sample statistics for each group.

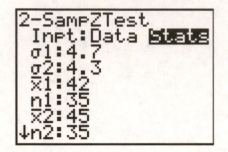

Select $\neq \mu_2$ as the alternative hypothesis and press **ENTER**. Scroll down to the next line and select **Calculate** or **Draw** and press **ENTER**.

The output for **Calculate** is:

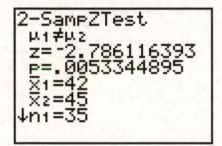

The output for **Draw** is:

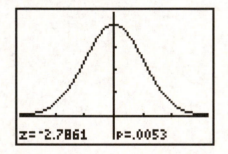

The test statistic, Z, is –2.786 and the P-value is .0053. Since the P-value is less than α, the correct decision is to **Reject H_o**. There is enough evidence at the 10% level of significance to support the safety engineer's claim that the mean braking distances are different for the two types of tires.   ◀

▶ Exercise 31 (pg. 437) Testing the Difference between 2 Population Means

Enter the data from 1981 into **L1** and enter the data from the more recent study into **L2**. The hypothesis test is: $H_o: \mu_1 \leq \mu_2$ vs.. $H_a: \mu_1 > \mu_2$. Since both n_1 and n_2 equal 30, the **Two-Sample Z-Test** can be used. In order to use this test, values for σ_1 and σ_2 are required. The sample standard deviations, s_1 and s_2 can be used as approximations for σ_1 and σ_2 when both n_1 and n_2 are greater than or equal to 30. To find the standard deviations, press **STAT**, highlight **CALC** and select **1:1-Var Stats** and press **2nd L1 ENTER**. The value for Sx is the sample standard deviation for the 1981 data. Next, press **STAT**, highlight **CALC** and select **1:1-Var Stats**, press **2nd L2 ENTER**. The value for Sx is the sample standard deviation for the more recent data.

To run the hypothesis, press **STAT**, highlight **TESTS** and select **3:2-SampZTest**. For **Inpt** select **Data** and press **ENTER**. Enter the sample standard deviations as approximations for σ_1 and σ_2. For **List1**, press **2nd [L1]** and for **List2** press **2nd [L2]**.

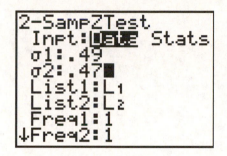

Select $> \mu_2$ as the alternative hypothesis and press **ENTER**. Scroll down to the next line and select **Calculate** or **Draw** and press **ENTER**.

The output for **Calculate** is:

```
2-SampZTest
 μ1>μ2
 z=3.011666878
 p=.001299157
 x̄1=2.13
 x̄2=1.756666667
↓Sx1=.490003519
```

The output for **Draw** is:

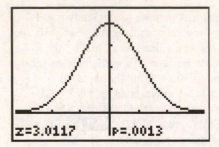

Since the P-value is smaller than α, the correct decision is to **Reject H₀.** There is enough evidence at the 2.5% level of significance to support the sociologist's claim that children ages 6 – 17 spent more time watching television in 1981 than children ages 6 – 17 do today.

▶ Exercise 45 (pg. 440) Confidence Interval for $\mu_1 - \mu_2$

Construct a confidence interval for $\mu_1 - \mu_2$, the difference between two population means. Designate the group using the DASH diet as population 1 and the group using the tradition diet as population 2. The sample statistics for Group1 are: $\bar{x}_1 = 123.1$mm, $s_1 = 9.9$mm and $n_1 = 269$. The sample statistics for Group2 are: $\bar{x}_2 = 125$mm, $s_2 = 10.1$mm and $n_2 = 268$.

Press **STAT**, highlight **TESTS** and select **9:2-SampZInt**. Since you are using the sample statistics for your analysis, select **Stats** for **Inpt** and press **ENTER**. Enter the sample standard deviations as approximations for σ_1 and σ_2. Enter the sample means and sample sizes for each group.

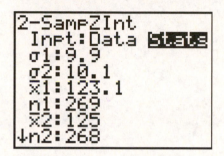

Scroll down to the next line and type in the confidence level of .95. Scroll down to the next line and press **ENTER**.

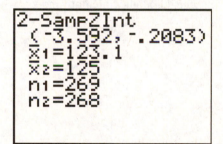

A 95% confidence interval for the difference in the population means is: (-3.592, -0.2083). The average systolic blood pressure for individuals on the DASH diet is between 0.2083 and 3.592 mm Hg lower than the average systolic blood pressure for individuals on the traditional diet and exercise plan. The DASH diet should be recommended over the traditional diet and exercise program because the mean systolic blood pressure is significantly lower on the DASH program.

◀

Section 8.2

▸ Example 1 (pg. 444) Two Sample t-Test

To test whether there is a difference in the mean mathematics test scores for the students of the two different teachers, use a two-tailed test: H_o: $\mu_1 = \mu_2$ vs.. H_a: $\mu_1 \neq \mu_2$.   The sample statistics are found in the table at the top of pg. 444 in your textbook.

Press **STAT**, highlight **TESTS**, and select **4:2-SampTTest**. Since you are inputting the sample statistics, select **Stats** and press **ENTER**. Enter the sample information from the two samples.

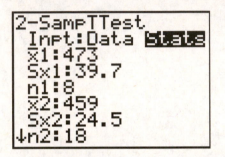

Scroll down to the next line and select $\neq \mu_2$ as the alternative hypothesis and press **ENTER**. Scroll down to the next line. On this line, you should select **NO** because the variances are NOT assumed to be equal and, therefore, you do not want a pooled estimate of the standard deviation. Press **ENTER**. On the next line, highlight **Calculate** and press **ENTER**.

```
2-SampTTest
 µ₁≠µ₂
 t=.9224141169
 P=.37924039
 df=9.458685946
 x̄₁=473
↓x̄₂=459
■
```

The output (shown above) for **Calculate** displays the alternative hypothesis, the test statistic, the P-value, the degrees of freedom and the sample statistics. Notice the degrees of freedom = 8.594. In cases, such as this one, in which the population variances are not assumed to be equal, the calculator calculates an adjusted degrees of freedom, rather than using the smaller of $(n_1 - 1)$ and $(n_2 - 1)$.

If you choose **Draw**, the output includes a graph with the area associated with the P-value shaded.

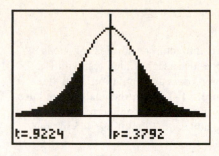

Both outputs display the P-value, which is 0.3792. Since the P-value is greater than α, the correct decision is to **Fail to Reject H₀.** At the 10% level of significance, there is not enough evidence to support the claim that there is a difference in the mean mathematics test scores for the students of the two different teachers.

◄

▶ Example 2 (pg. 445) Two Sample t-Test

To test the claim that the mean calling range of the manufacturer's cordless phone is greater than the mean range of the competitor's phone, use a right-tailed test with H_o: $\mu_1 \le \mu_2$ vs. H_a: $\mu_1 > \mu_2$. Designate the manufacturer's data as Population1 and the competitor's data as Population2. Use the **Two-Sample T-Test** for the analysis, and use the pooled variance option because the population variances are assumed to be equal in this example. The sample statistics are found in the table at the top of pg. 445 in your textbook.

Press **STAT**, highlight **TESTS**, and select **4:2-SampTTest**. Fill in the input screen with the sample statistics.

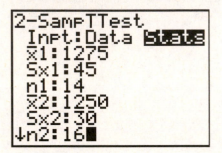

Select $> \mu_2$ for the alternative hypothesis and press **ENTER**. Select **YES** for **Pooled** and press **ENTER**. To do the analysis, select **Calculate** and press **ENTER**.

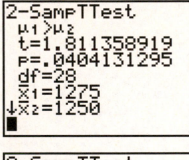

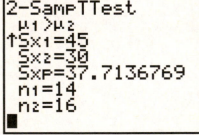

The two-page output display (shown above) includes the alternative hypothesis, the test statistic, the P-value, degrees of freedom, the sample statistics and the pooled standard deviation (Sxp = 37.7)

If you select **Draw**, the display includes a graph with the shaded area associated with the test statistic, the P-value and the test statistic.

Since the P-value is less than α, the correct decision is to **Reject H_0.** There is enough evidence at the 5% level of significance to support the manufacturer's claim that its phones have a greater calling range than its competitor's phones.

◄

▸ Exercise 16 (pg. 447) Testing the Difference Between 2 Population Means

Designate the small pick-ups as Population1 and the small SUVs. as Population2. Test the hypothesis: H$_o$: $\mu_1 = \mu_2$ vs.. H$_a$: $\mu_1 \neq \mu_2$. The sample statistics are: $\bar{x}_1 = 11.18$ cm, $s_1 = 4.53$ cm, $n_1 = 7$, $\bar{x}_2 = 9.52$ cm, $s_2 = 3.84$ cm and $n_2 = 13$. Use the **Two-Sample T-test** with the pooled variance option. Press **STAT**, highlight **TESTS** and select **4:2-SampTTest**. Choose **Stats** for **Inpt** and press **ENTER**. Enter the sample statistics. Select $\neq$ μ_2 as the alternative hypothesis and press **ENTER**. Select **YES** for **Pooled** and press **ENTER**. Highlight **Calculate** and press **ENTER**.

```
2-SampTTest
 μ1≠μ2
 t=.8672351426
 P=.3972274009
 df=18
 x̄1=11.18
↓x̄2=9.52
```

```
2-SampTTest
 μ1≠μ2
↑Sx1=4.53
 Sx2=3.84
 SxP=4.08297686
 n1=7
 n2=13
■
```

Notice that the test statistic = 0.8672, and the P-value is 0.3972. Since the P-value is greater than α, the correct decision is to **Fail to Reject H$_o$.** There is not enough evidence at the 1% level of significance to reject the claim that the mean footwell intrusions for small pick-ups and small SUVs. are equal.

◂

▶ Exercise 21(pg. 449) Testing the Difference between 2 Population Means

Designate the "Old Curriculum" as Population1 and the "New Curriculum" as Population2 and test the hypothesis: H_o: $\mu_1 \geq \mu_2$ vs. H_a: $\mu_1 < \mu_2$. Press **STAT**, select **1:Edit** and enter the data sets into **L1** (Old Curriculum) and **L2** (New Curriculum). Press **STAT**, highlight **TESTS** and select **4:2-SampTTest**. For this analysis, you are using the actual data so select **Data** for **Inpt** and press **ENTER**. Fill in the input screen with the appropriate information. Choose $< \mu_2$ as the alternative hypothesis and press **ENTER**. Choose **YES** for **Pooled** and press **ENTER**. Highlight **Calculate** and press **ENTER**.

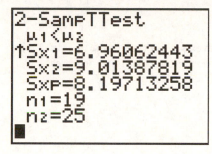

```
2-SampTTest
 μ1<μ2
 t=-4.29519297
 P=5.0503069E-5
 df=42
 x̄1=56.68421053
↓x̄2=67.4
```

```
2-SampTTest
 μ1<μ2
↑Sx1=6.96062443
 Sx2=9.01387819
 Sxp=8.19713258
 n1=19
 n2=25
```

Since the P-value (.00005) is less than α, the correct decision is to **Reject H_o.** There is enough evidence at the 10% level of significance to support the claim that using the New Curriculum results in a higher mean reading test score. The recommendation is to change to the new method.

◀

▸ Exercise 27 (pg. 450) Confidence Interval for $\mu_1 - \mu_2$

Compare the mean time spent waiting for a kidney transplant for two different age groups using a confidence interval. For this exercise, assume that the populations are approximately normal with equal variances. Also, notice that the sample sizes are both less than 30. The appropriate confidence interval technique is the **Two-Sample T-interval**.

Press **STAT**, highlight **TESTS** and select **0:2-SampTInt**. For **Inpt**, select **Stats** and press **ENTER**. Enter the sample statistics from the two sets of data. Designate the 35 – 49 Age Group as Population1 and the 50 – 69 Age Group as Population2.

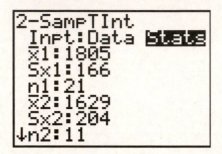

Scroll down to the next line and choose **.95** for **C-level,** scroll down to the next line and choose **YES** for **Pooled** and press **ENTER**. Scroll down to the next line and press **ENTER** to **Calculate** the confidence interval.

The output displays the 95% confidence interval for ($\mu_1 - \mu_2$) and the sample statistics. Based on a 95% confidence interval, the difference between mean waiting times for the two age groups is between 40 and 312 days. This means that the mean waiting time for the 35 – 49 Age Group (μ_1) is between 40 and 312 days longer than the mean waiting time for the 50 – 64 Age Group (μ_2). Note: The answer obtained using the TI-83/84 Plus differs slightly from the answer obtained using the confidence interval formula and the t-table. Consequently, your answer may not be exactly the same as the answer found in the answer key.

◂

▶ Exercise 29 (pg. 450) Confidence Interval for $\mu_1 - \mu_2$

Compare the mean driving distance of two golfers using a confidence interval. For this exercise, assume the populations are normal but do NOT assume equal variances.

Press **STAT**, highlight **TESTS** and select **0:2-SampTInt**. For **Inpt**, select **Stats** and press **ENTER**. Enter the sample statistics from the table displayed in Exercise 29.

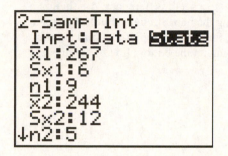

Scroll down to the next line and choose **.90** for **C-level**, scroll down to the next line and choose **NO** for **Pooled** and press **ENTER**. Scroll down to the next line and press **ENTER** to **Calculate** the confidence interval.

```
2-SampTInt
 (11.528,34.472)
 df=5.138708445
 x̄₁=267
 x̄₂=244
 Sx₁=6
↓Sx₂=12
```

 The output displays the 90% confidence interval for ($\mu_1 - \mu_2$), the adjusted degrees of freedom and the sample statistics. Based on a 90% confidence interval, the difference between mean driving distances for the two golfers is between 12 and 34 yards. This means that Golfer1 has a mean driving distance between 12 and 34 yds. longer than the mean driving distance for Golfer2. Note: The answer obtained using the TI-83/84 Plus differs slightly from the answer obtained using the confidence interval formula and the t-table. Consequently, your answer may not be exactly the same as the answer found in the answer key.

◀

Section 8.3

▸ Example 1 (pg. 453)   t-Test for the Difference Between Means (Dependent Samples)

In this example, the data is paired data, with two measurements of vertical jump heights for each of the 8 athletes. Enter the vertical jump heights before using the new Strength shoes for each athlete into L1. Enter the vertical jump heights using the new Strength shoes for each athlete into L2. Next, you must create a set of differences, d = (height before using the new Strength shoe) - (height using the new Strength shoe). To create this set, move the cursor to highlight the label **L3,** found at the top of the third column, and press **ENTER**. Notice that the cursor is flashing on the bottom line of the display. Press **2**nd **[L1]** - **2**nd **[L2]**

```
L1        L2        L3       3
24        26        ------
22        25
25        25
28        29
35        33
32        34
30        35

L3 =L1-L2
```

and press **ENTER**.

```
L1        L2        L3       3
24        26        -2
22        25        -3
25        25        0
28        29        -1
35        33        2
32        34        -2
30        35        -5

L3(1)= -2
```

Each value in **L3** is the difference **L1 - L2**.

To test the claim that athletes can increase their vertical jump heights with the new Strength shoes, the hypothesis test is: H_o: $\mu_d \geq 0$ vs.. H_a: $\mu_d < 0$. Press **STAT**, highlight **TESTS** and select **2:T-Test**. In this example, you are using the actual data to do the analysis, so select **Data** for **Inpt** and press **ENTER**. The value for μ_o is **0**, the value in the null hypothesis. The set of differences is found in **L3**, so set **List** to **L3**. Set **Freq** equal to **1**. Choose $< \mu_o$ as the alternative hypothesis and highlight **Calculate** or **Draw** and press **ENTER**.

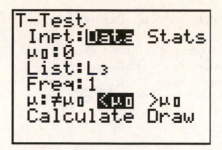

The output for **Calculate** is:

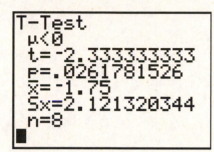

The output for **Draw** is:

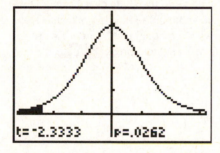

Since the P-value is less than α, the correct decision is to **Reject H_o**. There is enough evidence at the 10% level of significance to support the shoe manufacturer's claim that athletes can increase their vertical jump heights with the new Strength shoes.

▶ Exercise 10 (pg. 456) t-Test for the Difference Between Means (Dependent Samples)

Press **STAT** and select **1:Edit.** Enter the students' scores on the first SAT into **L1** and their scores on the second Sat into **L2**. Since this is paired data, create a column of differences, d = (first SAT) - (second SAT). To create the differences, move the cursor to highlight the label **L3,** found at the top of the third column, and press **ENTER**. Notice that the cursor is flashing on the bottom line of the display. Press **2nd [L1] - 2nd [L2]** and press **ENTER**.

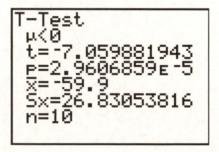

Each value in **L3** is the difference **L1 - L2**.

To test the claim that an SAT prep course improves the test scores of students, the hypotheses are: H_o: $\mu_d \geq 0$ vs. H_a: $\mu_d < 0$. Press **STAT**, highlight **TESTS** and select **2:T-Test**. In this example, you are using the actual data to do the analysis, so select **Data** for **Inpt** and press **ENTER**. The value for μ_o is **0**, the value in the null hypothesis. The set of differences is found in **L3**, so set **List** to **L3**. Set **Freq** equal to **1**. Choose < μ_o as the alternative hypothesis and highlight **Calculate** or **Draw** and press **ENTER**.

The output for **Calculate** is:

```
T-Test
 μ<0
 t=-7.059881943
 p=2.9606859ᴇ-5
 x̄=-59.9
 Sx=26.83053816
 n=10
```

The output for **Draw** is:

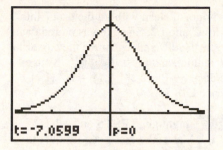

Since the P-value is less than α, the correct decision is to **Reject H$_o$**. There is enough evidence at the 1% level of significance to support the claim that the SAT prep course improves the test scores of students.

◄

▶ Exercise 23 (pg. 460)   Confidence Interval for μ_d

Press **STAT** and select **1:Edit**. Enter the data for "hours of sleep without the drug" into **L1** and the data for "hours of sleep using the new drug" into **L2**. Since this is paired data, create a column of differences, **d = L1 - L2**. To create the differences, move the cursor to highlight the label **L3,** found at the top of the third column, and press **ENTER**. Notice that the cursor is flashing on the bottom line of the display. Press **2nd [L1]** - **2nd [L2]** and press **ENTER**. Each value in **L3** is the difference **L1 - L2**.

To construct a confidence interval for μ_d, press **STAT**, highlight **TESTS** and select **8:Tinterval**. For **Inpt**, select **Data,** for **List**, enter **L3**. Make sure that **Freq** is equal to **1,** and the **C-level** is **90.** Highlight **Calculate** and press **ENTER**.

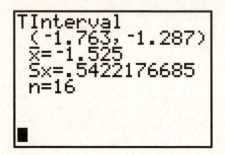

The confidence interval for μ_d is -1.763 to -1.287. This means that the average difference in hours of sleep is 1.287 hours to 1.763 hours more for patients using the new drug.

◀

Section 8.4

▶ Example 1 (pg. 463) Testing the Difference Between p_1 and p_2

To test the claim that the proportion of occupants who wear seat belts is the same for passenger cars and pickup trucks, the correct hypothesis test is: $H_o: p_1 = p_2$ vs. $H_o: p_1 \neq p_2$. Designate the Passenger Cars as Population1 and the Pickup Trucks as Population2. The sample statistics are $n_1 = 150$, $\hat{p}_1 = 0.86$, $n_2 = 200$, and $\hat{p}_2 = 0.74$. To conduct this test using the TI-83/84 Plus, you need values for x_1, the number of occupants in the sample of passenger cars who wore seatbelts, and x_2, the number of occupants of pickup trucks in the sample who wore seatbelts. To calculate x_1, multiply n_1 times $\hat{p}_1$. To calculate x_2 multiply n_2 times $\hat{p}_2$. (Note: These two values, x_1 and x_2, are given in the table on pg. 463.)

Press **STAT**, highlight **TESTS** and select **6:2-PropZTest** and fill in the appropriate information.

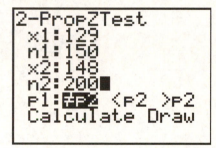

Highlight **Calculate** and press **ENTER**.

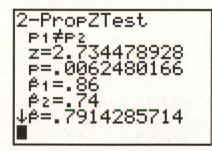

The output displays the alternative hypothesis, the test statistic, the P-value, the sample statistics and the weighted estimate of the population proportion, $\hat{p}$. Note: At this point in the analysis, you should confirm that the following products are each equal to 5 or more: $n_1\bar{p}$, $n_2\bar{q}$, $n_2\bar{p}$, and $n_2\bar{q}$. The value for $\bar{p}$ can be found in the above output, $\bar{p} = \hat{p}$. And $\bar{q} = 1 - \bar{p}$.

Since the P-value is less than α, the correct decision is to **Reject H_o**. There is enough evidence at the 10% level of significance to reject the claim the proportion of occupants who wear seat belts is the same for passenger cars and pickup trucks.

◀

Exercise 9 (pg. 465) The Difference Between Two Proportions

Designate subjects who wore magnetic insoles as Population1and subjects who wear nonmagnetic insoles as Population2 and test the hypotheses: H_o: p_1 = p_2 vs. H_o: $p_1 \neq$ p_2. The sample statistics are $x_1 = 17$, $n_1 = 54$, $x_2 = 18$ and $n_2 = 41$.

Press **STAT**, highlight **TESTS** and select **6:2-PropZTest** and fill in the appropriate information.

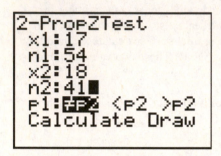

Highlight **Calculate** and press **ENTER**.

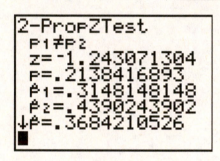

Note: At this point in the analysis, you should confirm that the following products are each equal to 5 or more: $n_1\bar{p}$, $n_2\bar{q}$, $n_2\bar{p}$, and $n_2\bar{q}$. The value for $\bar{p}$ can be found in the above output, $\bar{p} = \hat{p}$, and $\bar{q} = 1 - \bar{p}$.

Since the P-value is greater than α, the correct decision is to **Fail to Reject H_o**. There is not enough evidence at the 1% level of significance to support the claim that there is a difference in the proportion of subjects who feel all or mostly better after 4 weeks between the subjects who used magnetic insoles and subjects who used nonmagnetic insoles.

◀

▶ Exercise 29 (pg. 468)   Confidence Interval for p_1 - p_2

Construct a confidence interval to compare the proportion of students who had planned to study education 10 years ago to the proportion currently planning on studying education. Designate the current survey results as Population1 and the survey results from 10 years ago as Population 2. The sample statistics are $n_1 = 10000$, $\hat{p}_1 = 0.07$, $n_2 = 8000$, and $\hat{p}_2 = 0.09$. To conduct this test using the TI-83/84 Plus, you need values for x_1, the number of students in the current sample of students who plan on studying education, and x_2, the number of students in the survey 10 years ago who planned on studying education. To calculate x_1, multiply n_1 times $\hat{p}_1$.   To calculate x_2 multiply n_2 times $\hat{p}_2$.

Press **STAT**, highlight **TESTS** and select **B:2-PropZInt** and fill in the appropriate information.

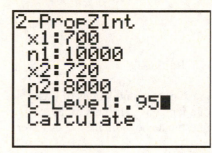

Highlight **Calculate** and press **ENTER**.

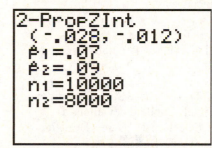

The confidence interval (-.028, -.012) indicates that the proportion of students currently choosing to study education between 1.2% and 2.8 % lower than the proportion of students who chose to study education 10 years ago.

◀

▶ Technology (pg. 477) Tails Over Heads

Exercise 1 - 2: Test the hypotheses: H_o: P(Heads) = .5 vs. H_a: P(Heads) $\neq$ .5 using the one sample test of a proportion. Press **STAT**, highlight **TESTS** and select **5:1-PropZTest.** For this example, p_o = .5. Using Casey's data, X = 5772 and n = 11902. The alternative hypothesis is $\neq$. Highlight **Calculate** and press **ENTER**.

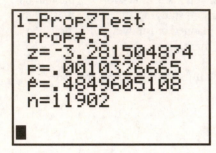

Since the P-value is less than α, the correct decision is to **Reject H_o**.

Exercise 3: The histogram at the top of the page is a graph of 500 simulations of Casey's experiment. Each simulation represents 11902 flips of a fair coin. The bars of the histogram represent frequencies. Use the histogram to estimate how often 5772 or fewer heads occurred.

To simulate this experiment, you must use an alternative technology. The TI-83/84 Plus does not have the memory capacity to do this experiment.

Exercise 4: To compare the mint dates of the coins, run the hypothesis test:
H_o: $\mu_1 = \mu_2$ vs.. H_a: $\mu_1 \neq \mu_2$. Designate the Philadelphia data as Population1 and the Denver data as Population 2.

Press **STAT**, highlight **TESTS** and select **3:2-SampZTest**. Choose **Stats** for **Inpt** and press **ENTER**. Enter the sample statistics. Use the sample standard deviations as approximations to the population standard deviations. Select $\neq \mu_2$ as the alternative hypothesis and press **ENTER**. Highlight **Calculate** and press **ENTER**.

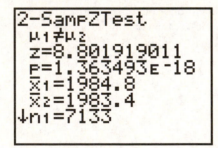

Since the P-value is extremely small, the correct decision is to **Reject H_o**.

Exercise 5: To compare the average mint value of coins found in Philadelphia to those found in Denver, run the hypothesis test: H_o: $\mu_1 = \mu_2$ vs. H_a: $\mu_1 \neq \mu_2$. Designate the Philadelphia data as Population1 and the Denver data as Population2.

Press **STAT**, highlight **TESTS** and select **3:2-SampZTest**. Choose **Stats** for **Inpt** and press **ENTER**. Enter the sample statistics. Use the sample standard deviations as approximations to the population standard deviations. Select $\neq \mu_2$ as the alternative hypothesis and press **ENTER**. Highlight **Calculate** and press **ENTER**. Since the P-value is greater than α, the correct decision is to **Fail to Reject H_o**.

◀

CHAPTER
9

Section 9.1

▸ Example 3 (pg. 486) Constructing a Scatter Plot

Press **STAT**, highlight **1:Edit** and clear **L1** and **L2**. Enter the X-values into **L1** and the Y-values into **L2**. To prepare to construct the scatterplot, press the **Y=** key and clear all the Y-registers. Press **2ⁿᵈ** **[STAT PLOT]**, select **1:Plot1**, turn **On** Plot 1 and press **ENTER**. For **Type** of graph, select the **scatterplot** which is the first selection. Press **ENTER**. Enter **L1** for **Xlist** and **L2** for **Ylist**. Highlight the first selection, the small square, for the type of **Mark**. Press **ENTER**. Press **ZOOM** and **9** to select **ZoomStat**.

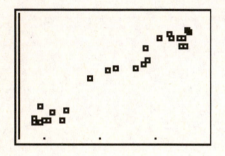

This graph shows a positive linear correlation.

◂

> ▸ Example 5 (pg. 489)   Finding a Correlation Coefficient

For this example, use the data from Example 3 on pg. 486. Enter the X-values into **L1** and the Y-values into **L2**. In order to calculate r, the correlation coefficient, you must turn **On** the **Diagnostic** command. Press **2ⁿᵈ** **[CATALOG]** (Note: CATALOG is found above the **0** key). The CATALOG of functions will appear on the screen. Use the down arrow to scroll to the **DiagnosticOn** command.

Press **ENTER** **ENTER**.

Press **STAT**, highlight **CALC**, scroll down to **4:LinReg(ax+b)** and press **ENTER** **ENTER**. (Note: This command allows you to specify which lists contain the X-values and Y-values. If you do not specify these lists, the defaults are used. The defaults are: **L1** for the X-values and **L2** for the Y-values.)

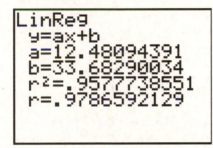

The correlation coefficient is r = .9786592129. This indicates a strong positive linear correlation between X and Y.

◂

▶ Exercise 23 (pg. 497) Constructing a Scatterplot and Determining r

Enter the X-values into **L1** and the Y-values into **L2**. To prepare to construct the
scatterplot, press the **Y=** key and clear all the Y-registers. Press **2ⁿᵈ** **[STAT PLOT]**,
select **1:Plot1**, turn **On** Plot 1 and press **ENTER**. For **Type** of graph, select the **scatter
plot**. Enter **L1** for **Xlist** and **L2** for **Ylist**. Press **ZOOM** and **9** to select **ZoomStat**.

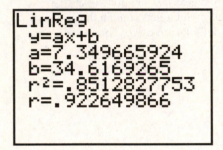

To calculate r, press **STAT**, highlight **CALC** and scroll down to **4:LinReg(ax+b)** and
press **ENTER ENTER**. (Note: If you have not turned **On** the **Diagnostics,** do that first
by following the instructions for Example 5 on the previous page.)

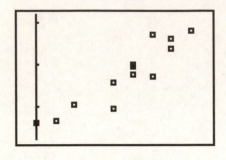

The scatter plot shows a strong positive linear correlation. This is confirmed by the
r-value of 0.923.

◀

▸ Exercise 33 (pg. 498)   Testing Claims

For this example, use the data from Exercise 23 on pg. 497. To test the significance of the population correlation coefficient, ρ, the appropriate hypothesis test is: $\rho = 0$ vs. $\rho \neq 0$. To run the test, enter the X-values into **L1** and the Y-values into **L2**. Press **STAT**, highlight **TESTS** and select **E:LinRegTTest**. Enter **L1** for **Xlist**, **L2** for **Ylist**, and **1** for **Freq**. On the next line, β and ρ, select $\neq 0$ and press **ENTER**. Leave the next line, RegEQ, blank. Highlight **Calculate.**

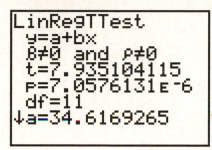

```
LinRegTTest
Xlist:L₁
Ylist:L₂
Freq:1
β & ρ:≠0 <0 >0
RegEQ:
Calculate
```

Press **ENTER**.

```
LinRegTTest
y=a+bx
β≠0 and ρ≠0
t=7.935104115
p=7.0576131ε⁻6
df=11
↓a=34.6169265
```

The output displays several pieces of information describing the relationship between X and Y. What you are interested in for this example are the following: the test statistic (t = 7.935), the P-value (p = 7.0576131E-6) and the r value (r = .922649886). Since the P-value is less than α, the correct decision is to **Reject** the null hypothesis. There is enough evidence at the 1% level of significance to conclude that there is a significant linear relationship between number of hours spent studying for a test and the score received on the test.

◀

Section 9.2

▸ Example 2 (pg. 503)   Finding a Regression Line

Enter the X-values into **L1** and the Y-values into **L2**. Press **STAT**, highlight **CALC** and scroll down to **4:LinReg(ax+b)**. This command has several options. One option allows you to store the regression equation into one of the Y-variables. To use this option, with the cursor flashing on the line **LinReg(ax+b)**, press **VARS**.

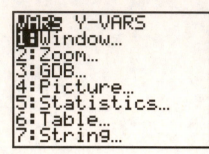

Highlight **Y-VARS**.

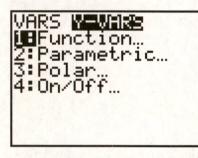

Select **1:Function** and press **ENTER**

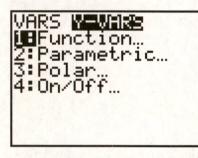

Notice that **1:Y1** is highlighted. Press **ENTER**.

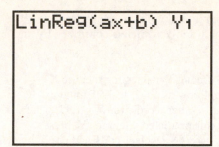

Press **ENTER**.

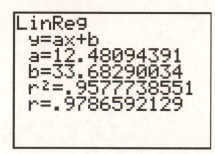

The output displays the general form of the regression equation: **y = ax+b** followed by values for **a** and **b**. On the next line you will see r^2, the coefficient of determination, followed by **r**, the correlation coefficient. If you put the values of **a** and **b** into the general equation, you obtain the specific linear equation for this data:
$\hat{y} = 12.48\,x + 33.68$. Press **Y=** and see that this specific equation has been pasted to **Y1**.

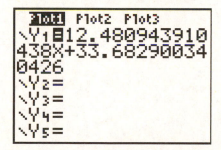

Press **2ⁿᵈ** **[STAT PLOT]**, select **1:Plot1**, turn **On** Plot1, select **scatterplot**, set **Xlist** to **L1** and **Ylist** to **L2**. Press **ZOOM** and **9**.

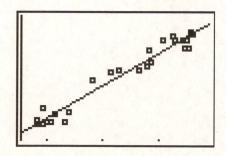

This picture displays a scatterplot of the data and the regression line. The picture indicates a strong positive linear correlation between X and Y, which is confirmed by the r-value of 0.979.

You can use the regression equation stored in **Y1** to predict Y-values for specific X-values. For example, suppose the duration of an eruption was equal to 1.95 minutes. Predict the time (in minutes) until the next eruption. In other words, for X = 1.95, what does the regression equation predict for Y? To find this predicted value for Y, press **VARS**, highlight **Y-VARS**, select **1:Function**, press **ENTER**, select **1:Y1** and press **ENTER**. Press **(** 1.95 **)** and press **ENTER** .

```
Y₁(1.95)
        58.02074097
```

The output is a display of the predicted Y-value for X = 1.95.

▶ Exercise 17 (pg. 506)   Finding the Equation of the Regression Line

Enter the X-values into **L1** and the Y-values into **L2**. Press **STAT**. Highlight **CALC**, select **4:LinReg(ax+b).** With the cursor flashing on the line **LinReg(ax+b)**, press **VARS**, highlight **Y-VARS**, select **1:Function**, press **ENTER** and select **1:Y1** and press **ENTER** **ENTER**.

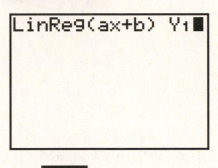

Press **ENTER**.

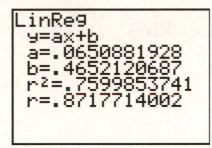

Using **a** and **b** from the output display, the resulting regression equation is $\hat{y} = 0.06509\,x + 0.4652$. Press **Y=** to confirm that the regression equation has been stored in **Y1**.

To view the data and the regression line, first make sure that the scatterplot has been selected. Press **2ⁿᵈ** **[STAT PLOT]**, select **1:Plot1**, turn **On** Plot1, select **scatterplot**, set **Xlist** to **L1** and **Ylist** to **L2**. Press **ZOOM** and **9** and a graph of the scatter plot with the regression line will be displayed.

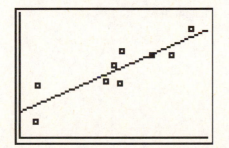

Next, you can use the regression equation to predict number of stories, Y, for various heights, X. First, check the X-values that you will be using to confirm that they are within (or close to) the range of the X-values in your data. Three X-values (800, 750 and

625) meet this criteria. The value of 400 is outside the range of the original X-values, so the regression equation will not be used with this value.

Press **VARS**, highlight **Y-VARS**, select **1:Function** and press **ENTER**. Select **1:Y1** and press **ENTER**. Press **(** 800 **)** and **ENTER**.

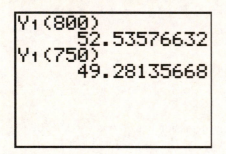

The predicted number of stories for a building with height of 500 ft. is approximately 53 stories.

Press **2ⁿᵈ** **[ENTRY]**, (found above the **ENTER** key). Move the cursor so that it is flashing on '**8**' in the number '800' and type in 750. Press **ENTER**.

```
Y₁(800)
        52.53576632
Y₁(750)
        49.28135668
```

The predicted number of stories for a building with a height of 750 ft. is 49 stories.

Press **2ⁿᵈ** **[ENTRY]**. Move the cursor so that it is flashing on '**7**' in the number '750' and type in 625. Press **ENTER**.

The predicted number of stories for a building with a height of 625 ft. is 42 stories.

◄

Section 9.3

▶ Example 2 (pg. 516)   The Standard Error of the Estimate

Enter the data into **L1** and **L2**. Press **STAT**, highlight **CALC**, select **4:LinReg(ax+b)**.
With the cursor flashing on the line **LinReg(ax+b)**, press **VARS**, highlight **Y-VARS**,
select **1:Function**, press **ENTER** and select **1:Y1** and press **ENTER** **ENTER**.

The formula for s_e, the standard error of the estimate is $\sqrt{\dfrac{\sum (y_i - \hat{y}_i)^2}{n-2}}$. The values

for $(y_i - \hat{y}_i)$ called Residuals, are automatically stored to a list called **RESID**. Press **2nd**
[LIST] (Note: **[LIST]** is found above the **STAT** key), select **RESID**. Press **STO** **2nd**
[L3] and **ENTER**. This stores the residuals to **L3**.

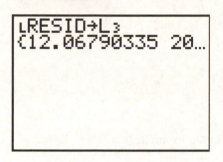

In the formula for s_e, the residuals, $(y_i - \hat{y}_i)$, are squared. To square these values and
store them in **L4**, press **STAT**, select **1:Edit** and move the cursor to highlight the
Listname **L4**. Press **ENTER**. Press **2nd** **[L3]** and the x^2 key.

L2	L3	L4	4
428.2	12.068	------	
828.8	20.364		
1214.2	150.77		
444.6	126.54		
264	-14.83		
415.3	-255.8		
571.8	-60.1		

L4 =L₃²

Press **ENTER**.

L2	L3	L4	4
428.2	12.068	145.63	
828.8	20.364	414.69	
1214.2	150.77	22731	
444.6	126.54	16013	
264	-14.83	219.8	
415.3	-255.8	65449	
571.8	-60.1	3611.9	

L4(1)=145.6342913...

Press **2ⁿᵈ** **[QUIT]**. Next, you need to find the sum of **L4**, $\sum (y_i - \hat{y}_i)^2$. Press **2ⁿᵈ** **[LIST]**. Highlight **MATH**, select **5:sum(** and press **2ⁿᵈ** **[L4]**. Close the parentheses and press **ENTER**.

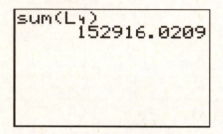

```
sum(L4)
        152916.0209
```

Divide this sum by (number of observations – 2). In this example, (n - 2) = 8. Simply press ÷ 8 and press **ENTER**.

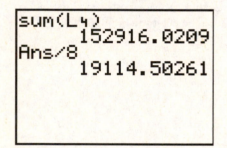

```
sum(L4)
        152916.0209
Ans/8
        19114.50261
```

Lastly, take the square root of this answer by pressing **2ⁿᵈ** **[√]** **2ⁿᵈ** **[ANS]**. Close the parentheses and press **ENTER**.

```
sum(L4)
        152916.0209
Ans/8
        19114.50261
√(Ans)
        138.2552083
```

The standard error of the estimate, s_e, is 138.26.

Example 3 (pg. 518)   Constructing a Prediction Interval

This example is a continuation of Example 2 on pg. 516. To construct a 95% prediction interval for a specific X-value, x_0, you must calculate the maximum error, E. The

formula for E is: $E = t_c s_e \sqrt{\left(1 + \dfrac{1}{n} + \dfrac{n(x_0 - \overline{x})^2}{n(\sum x^2) - (\sum x)^2}\right)}$. The critical value for t is

found using the inverse-t distribution function on the TI-84 calculator. (Note: If you are using a TI-83 calculator, you cannot find the t-value on the calculator. Use a t-table to find the t-value.) On the TI-84, press **2ⁿᵈ [DISTR]** and select **4:invT.** Type in the cumulative area for the upper end of the 95 % confidence interval: .975, and the degrees of freedom: $n - 1 = 8$. Press ENTER.

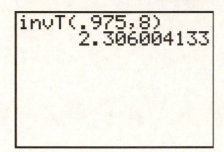

```
invT(.975,8)
        2.306004133
```

For this example, t_c = 2.306. The standard error, s_e, is 138.26. (Note: The standard error was calculated in Example 2 on the previous pages.)

To calculate $\overline{x}$, $\sum x^2$, and $(\sum x)^2$, press VARS, select **5:Statistics**. Highlight **2:** $\overline{x}$ and press ENTER ENTER. Notice that $\overline{x} = 2.31$. Press VARS again, select **5:Statistics**, highlight $\sum$, select **1:** $\sum x$ and press ENTER ENTER. So, $\sum x =$ 23.1. Press VARS again, select **5:Statistics**, highlight $\sum$, select **2:** $\sum x^2$, and press ENTER ENTER. Notice that $\sum x^2 = 67.35$.

Next, calculate the maximum error, E, when $x_0 = 3.5$ using the formula for E.

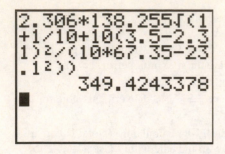

The prediction interval is $\hat{y} \pm 349.424$. To find $\hat{y}$, the predicted value for y when x = 3.5, press **VARS**, highlight **Y-VARS**, select **1:Function** and press **ENTER**. Next select **1:Y1**, press **ENTER** and press **(** 3.5 **)** . Finally, press **ENTER**.

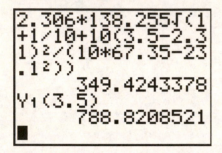

The prediction interval is 788.82 ± 349.42.

◀

▶ Exercise 13 (pg. 520)   Coefficient of Determination and Standard Error of the
                                        Estimate

Enter the data into **L1** and **L2**. Press **STAT**, highlight **CALC**, select **4:LinReg(ax+b).**
With the cursor flashing on the line **LinReg(ax+b)**, press **VARS**, highlight **Y-VARS**,
select **1:Function**, press **ENTER** and select **1:Y1** and press **ENTER ENTER**.

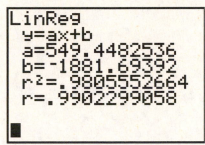

The coefficient of determination, r^2, is .9806. This means that 98.06 % of the variation
in the Y-values is explained by the X-values.

To calculate s_e, press **2nd [LIST]** , select **RESID**, press **ENTER**, **STO** , **2nd [L3]**
ENTER. This stores the residuals to **L3**. In the formula for s_e, the residuals, $(y_i - \hat{y}_i)$
are squared. To square these values and store them in **L4**, press **STAT**, select **1:Edit** and
move the cursor to highlight the Listname **L4**. Press **ENTER**. Press **2nd [L3]** and the x^2
key. Press **ENTER**.

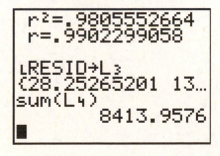

Press **2nd [QUIT]** . Next, you need to find the sum of **L4**, $\sum (y_i - \hat{y}_i)^2$. Press **2nd**
[LIST] . Highlight **MATH**, select **5:sum(** and press **2nd [L4]**. Close the parentheses
and press **ENTER**.

Divide this sum by (number of observations – 2). In this example, (n - 2) = 9. Simply press ÷ 9 and press **ENTER**.

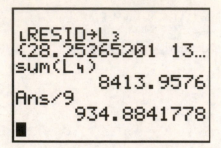

Lastly, take the square root of this answer by pressing **2nd** [√] **2nd** [ANS] . Close the parentheses and press **ENTER**.

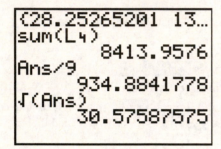

The standard error of the estimate is $30.576 billion.

Technology (pg. 537)   Nutrients in Breakfast Cereals

Exercises 1-2: Enter the data into **L1**, **L2**, **L3** and **L4**. To prepare to construct the scatterplots, press the **Y=** key and clear all the Y-registers. To construct the scatter plots, press **2nd** **[STAT PLOT]**, select **1:Plot1** and press **ENTER**. Turn **ON** Plot1. Select **scatter plot** for **Type**. Enter the appropriate labels for **Xlist** and **Ylist** to construct each of the scatter plots.

Exercises 3: To find the correlation coefficients, press **STAT**, highlight **CALC** and select **4:LinReg(ax+b)** and enter the labels of the columns you are using for the correlation. For example, to find the correlation coefficient for "fat" and "carbohydrates", press **2nd** **[L3]** **[,]** **2nd** **[L4]**.

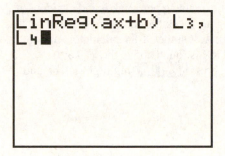

Press **ENTER** and the correlation coefficient for fat and carbohydrates is displayed in the output.

Exercise 4: To find the regression equations, press **STAT**, highlight **CALC** and select **4:LinReg(ax+b)** and enter the labels of the columns you are using for the regression. For example, to find the regression equation for "calories" and "carbohydrates", press **2nd** **[L1]** **[,]** **2nd** **[L4]** .

Exercise 5: Use the regression equations found in Exercise 4 to do the predictions.

Exercises 6 and 7: The TI-83/84 does not do multiple regression.

CHAPTER 10

Chi-Square Tests and the F-Distribution

Section 10.2

▶ Example 3 (pg. 556) Chi-Square Independence Test

Test the hypotheses: H_o: The number of days spent exercising per week is independent of gender vs. H_a: The number of days spent exercising per week depends on gender. The first step is to enter the data in the table into **Matrix A**. On the TI-84 calculators and the TI-83 Plus calculator, press 2nd **[MATRX]**. (MATRX is found above the x^{-1} key.) Highlight **EDIT** and press **ENTER**. On the TI-83, press **MATRX**, highlight **EDIT** and press **ENTER**.

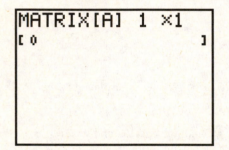

On the top row of the display, enter the size of the matrix. The matrix has 2 rows and 4 columns, so press **2**, press the right arrow key, and press **4**. Press **ENTER**. Enter the first value, 40, and press **ENTER**. Enter the second value, 53, and press **ENTER**. Continue this process and fill the matrix.

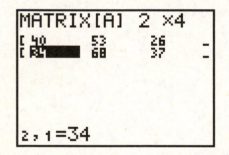

Press 2nd **[Quit]**. To perform the test of independence, press **STAT**, highlight **TESTS**, and select **C: χ^2-Test** and press **ENTER**.

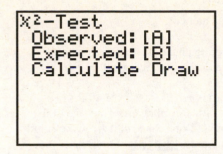

For **Observed**, **[A]** should be selected. If **[A]** is not already selected, press **2ⁿᵈ** **[MATRX]**, highlight **NAMES**, select **1:[A]** and press **ENTER**. For, **Expected**, **[B]** should be selected. Move the cursor to the next line and select **Calculate** and press **ENTER**.

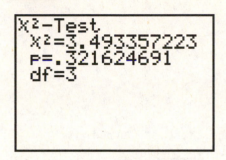

The output displays the test statistic and the P-value. Since the P-value is greater than α, the correct decision is to **Fail to Reject** the null hypothesis. This means that there is not enough evidence to conclude that the number of days spent exercising per week is related to gender.

Or, you could highlight **Draw** and press **ENTER**.

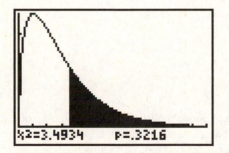

This output displays the χ^2 **–curve** with the area associated with the P-value shaded in. The test statistic and the P-value are also displayed.

▸ Exercise 17 (pg. 559)   Chi-Square Test for Independence

Test the hypotheses: H_o: The Result (improvement or no change) is independent of Treatment (drug or placebo) vs. H_a: The Result depends on the Treatment.

The first step is to enter the data in the table into **Matrix A**. Press **2ⁿᵈ [MATRX]**, highlight **EDIT** and press **ENTER**. On the top row of the display, enter the size of the matrix. The matrix has 2 rows and 2 columns, so press **2**, press the right arrow key, and press **2**. Press **ENTER**. Enter the first value, 39, and press **ENTER**. Enter the second value, 25, and press **ENTER**. Continue this process and fill the matrix.

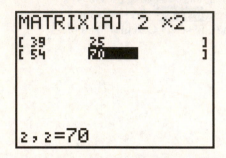

Press **2ⁿᵈ [Quit]**. Press **STAT**, highlight **TESTS**, and select **C: χ^2-Test** and press **ENTER**. For **Observed**, **[A]** should be selected. If **[A]** is not already selected, press **2ⁿᵈ [MATRX]**, highlight **NAMES**, select **1:[A]** and press **ENTER**. For, **Expected**, **[B]** should be selected. Move the cursor to the next line and select **Calculate** and press **ENTER**.

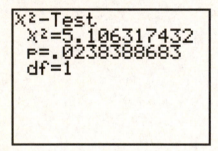

The output displays the test statistic and the P-value. Since the P-value is less than α, the correct decision is to **Reject** the null hypothesis. There is enough evidence at the 10% level of significance to conclude that results are dependent on type of treatment. The correct recommendation is to use the drug as part of the treatment.

◂

Section 10.3

> ► Example 3 (pg. 569)   Performing a Two-Sample F-Test

Test the hypotheses: H_o: $\sigma_1^2 \leq \sigma_2^2$ vs. H_o: $\sigma_1^2 > \sigma_2^2$. The sample statistics are: $s_1^2 = 400$, $n_1 = 10$, $s_2^2 = 256$ and $n_2 = 21$. Press **STAT**, highlight **TESTS** and select **2-SampFTest**. (Note: **2-SampFTest** is selection **E:** on the TI-84 and selection **D:** on the TI-83.) For **Inpt**, select **Stats** and press **ENTER**. On the next line, enter s_1, the standard deviation under the old system. The standard deviation is the square root of the variance, so press **2nd** [$\sqrt{\;}$] and enter 400. Enter n_1 on the next line. Next, enter the standard deviation under the new system by pressing **2nd** [$\sqrt{\;}$] 256. Enter n_2 on the next line. Highlight $> \sigma_2$ for the alternative hypothesis and press **ENTER**. Select **Calculate** and press **ENTER**.

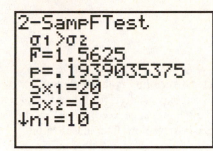

The output displays the alternative hypothesis, the test statistic, the P-value and the sample statistics. Since the P-value is greater than α, the correct decision is to **Fail to Reject** the null hypothesis. There is not enough evidence to conclude that the new system decreased the variance in waiting times.

You could also select **Draw** and press **ENTER**.

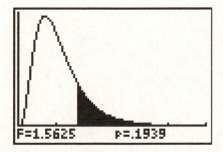

The output displays the F-distribution with the area associated with the P-value shaded in.

◄

▶ Exercise 25 (pg. 572)   Comparing Two Variances

Test the hypotheses: H_o: $\sigma_1^2 \leq \sigma_2^2$ vs. H_o: $\sigma_1^2 > \sigma_2^2$. The sample statistics are: $s_1 =$ $14,900, $n_1 = 16$, $s_2 = $9,600$ and $n_2 = 17$. Press **STAT**, highlight **TESTS** and select **2-SampFTest**. For **Inpt**, select **Stats** and press **ENTER**. On the next line, enter s_1, the standard deviation for annual actuarial salaries in New York. Enter n_1 on the next line. Next, enter the standard deviation and sample size for the annual actuarial salaries in California and press **ENTER**. Highlight $> \sigma_2$ for the alternative hypothesis and press **ENTER**. Select **Calculate** and press **ENTER**.

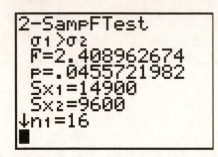

The output displays the alternative hypothesis, the test statistic, the P-value and the sample statistics. Since the P-value is less than α, the correct decision is to **Reject** the null hypothesis. There is enough evidence at the 5% level of significance to conclude that the standard deviation of annual actuarial salaries is greater in New York than in California.

◀

Section 10.4

▸ Example 2 (pg. 579)   Performing an ANOVA Test

Test the hypotheses: H_o: $\mu_1 = \mu_2 = \mu_3$ vs. H_a: at least one mean is different from the others. Enter the data into **L1, L2**, and **L3**. Press **STAT**, highlight **TESTS** and select **H:ANOVA(** and press **ENTER**, **2ⁿᵈ [L1]** , **2ⁿᵈ [L2]** , **2ⁿᵈ [L3].**

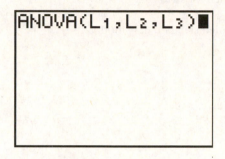

Press **ENTER** and the results will be displayed on the screen.

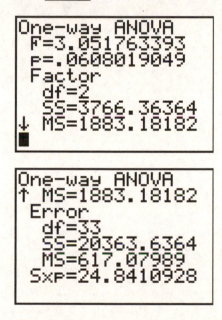

The output displays the test statistic, F = 3.05, and the P-value, p = .0608. Since the P-value is less than α, the correct decision is to **Reject** the null hypothesis. There is enough evidence at the 10% level of significance to reject the claim that the mean earnings are the same.

The output also displays all the information that is needed to set up an ANOVA table like the table on pg. 576 of your textbook.

Variation	Sum of Squares	Degrees of Freedom	Mean Squares	F
Between	3766.36	2	1883.18	1883.18/617.08
Within	20363.64	33	617.08	

Note: The TI-83/84 Plus labels variation "Between" samples as variation due to the "Factor". Also, variation "Within" samples is labeled as "Error".

The final item in the TI83/84 Plus output is the pooled standard deviation, Sxp = 24.84.

◀

▶ Exercise 5 (pg. 581)   Performing an ANOVA Test

Test the hypotheses: $H_o: \mu_1 = \mu_2 = \mu_3$ vs. H_a: at least one mean is different from the others. Enter the data into **L1, L2,** and **L3**. Press **STAT**, highlight **TESTS** and select **H:ANOVA(** and press **ENTER**, 2nd **[L1]** , 2nd **[L2]** , 2nd **[L3].** Press **ENTER** and the results will be displayed on the screen.

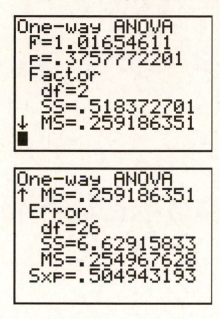

The output displays the test statistic, F = 1.017, and the P-value, p = 0.376. Since the P-value is greater than α, the correct decision is to **Fail to Reject** the null hypothesis. At the 5% level of significance, the data does not support the claim that at least one of the mean costs per ounce is different.

◀

▶ Exercise 13 (pg. 584)   Performing an ANOVA Test

Test the hypotheses: H_o: $\mu_1 = \mu_2 = \mu_3 = \mu_4$ vs. H_a: at least one mean is different from the others. Enter the data into **L1, L2, L3** and **L4**. Press **STAT**, highlight **TESTS** and select **F:ANOVA(** and press **ENTER**, 2nd **[L1]** **,** 2nd **[L2]** **,** 2nd **[L3]** **,** 2nd **[L4].** Press **ENTER** and the results will be displayed on the screen.

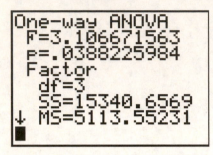

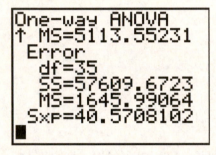

The output displays the test statistic, F = 3.107, and the P-value, p = .0388. Since the P-value is less than α, the correct decision is to **Reject** the null hypothesis. There is enough evidence at the 10% level of significance to conclude that the mean energy consumption for at least one region is different.

◀

▶ Technology (pg. 595)   Teacher Salaries

Exercise 2: Enter salaries for the 3 states into L1, L2 and L3. One method of testing to see if a set of data is approximately normal is to use a Normal Probability Plot. To prepare to graph the Normal Probability Plot, clear all Y-registers. To test the data in **L1**, press 2nd **[STAT PLOT]**, select Plot 1 and turn **ON** Plot1. For **Type**, select the last icon on the second line. For **Data List,** press 2nd **[L1]**. For **Data Axis**, select **X**. For **Mark**, select the first icon. Press **ZOOM** and **9** to display the normal plot.

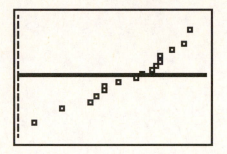

Data that is approximately normally distributed will have a plot that looks fairly linear. Although this graph is not perfectly straight, it is fairly linear and it is therefore reasonable to conclude that the data is approximately normal. Repeat this process to check the other datasets for normality.

Exercise 3: Test the hypotheses: H_o: $\sigma_1^2 = \sigma_2^2$ vs. H_a: $\sigma_1^2 \neq \sigma_2^2$ for each pair of samples. This test requires values for s_1 and s_2. To find the standard deviation for **L1**, press **STAT**, highlight **CALC** and press **L1 ENTER**. To find the standard deviation for **L2**, press **STAT**, highlight **CALC**, press **[L2] ENTER**. Repeat this process to get the standard deviation of **L3**. For each comparison, press **STAT**, highlight **TESTS** and select **D:2-SampFTest**. For **Inpt**, select **Stats** and press **ENTER**. To compare σ_1^2 and σ_2^2, enter s_1, the sample standard deviation for sample1. Enter n_1, the corresponding sample size, on the next line. Next, enter s_2 and n_2. Highlight $\neq \sigma_1^2$ for the alternative hypothesis and press **ENTER**. Select **Calculate** and press **ENTER**. Repeat this process to compare σ_1^2 and σ_3^2 using the sample standard deviations for L1 and L3. Finally compare σ_2^2 and σ_3^2 using the sample standard deviations for L2 and L3.

Exercise 4: Test the hypotheses: H_o: $\mu_1 = \mu_2 = \mu_3$ vs. H_a: at least one mean is different from the others. Press **STAT**, highlight **TESTS** and select **F:ANOVA(** and press **ENTER**, 2nd **[L1]** **,** 2nd **[L2]** **,** 2nd **[L3]**. Press **ENTER** and the results will be displayed on the screen.

Exercise 5: Repeat Exercises 1 - 4 using this second set of data.